mathematics of decision theory

methods and models
in the social sciences

3

MOUTON · THE HAGUE · PARIS

mathematics of decision theory

by PETER C. FISHBURN

MOUTON · THE HAGUE · PARIS

Library of Congress Catalog Card Number: 75-171099

Jacket-design: Jurriaan Schrofer
Printed in Hungary

To Jan and our daughters Susan, Kathy, and Sally

contents

preface

This little book describes the mathematics used in axiomatic investigations in decision theory. Its purpose is to introduce the student to these areas of mathematics and to show how they can apply to the analysis of specific types of decision situations, including multiple-factor situations and situations that involve uncertainty. Social choice or group decision is considered along with individual decision.

The book is organized by specific mathematical topics. Following an introduction that includes some elementary theory of sets and functions there are two chapters on binary relations and choice functions, upon which almost all decision theories are built. These are followed by a chapter that shows how the theory of the solution of finite systems of linear inequalities applies to axiomatic investigations in virtually every area of decision theory.

Our attention then turns to mathematics that are mainly concerned with the treatment of infinite sets. The first of these chapters shows how Zorn's lemma applies to several specific questions in decision theory: the second presents Cantor's Theorem on the existence of order-preserving real-valued functions. The final four chapters are based on structures of sets, or sets of sets, that are closed under one or more operations. Included are topologies, algebras of sets (used in probability theory), mixture sets, and ordered groups and groupoids. In each of these areas we shall note some of the special characteristics of decision theories that arise from the special structures.

I am indebted to many friends for their help over the years. My greatest debt is to Professor Leonard J. Savage.

I extend my sincerest thanks to Mrs. Virginia M. Johnson for her superb typing of the manuscript.

Potomac, Maryland
January 1970

Peter C. Fishburn

CHAPTER 1

mathematics of decision theory

This book discusses the mathematics of axiomatic decision theories. Such a theory consists of a set of axioms (assumptions, postulates, hypotheses, conditions, criteria) and the conclusions (theorems, lemmas, corollaries) that can be deduced from the axioms. The axioms usually refer to preference or choice structures of an individual or a group of individuals.

One conclusion that might follow from a set of axioms is that the axioms are logically inconsistent. This occurs for example when some of the axioms imply the contradictory of another axiom in the set. Although our primary concern is with consistent theories, the importance of inconsistent axiom sets will be noted.

A simple though important example of a consistent theory goes as follows. An individual must select one alternative from a finite set of decision alternatives. It is assumed that he prefers no alternative to itself, and if he prefers one alternative to a second and the second to a third then he prefers the first to the third. It then follows that a number can be assigned to each alternanative so that the number assigned to one alternative exceeds the number assigned to a second whenever the individual prefers the first alternative to the second.

The mathematics of this simple case do not depend on the notions of individual, alternative, and preference. These terms provide only an extra-mathematical interpretation of the theory. When stripped of this interpretation, what remains can be written as follows:

AXIOM A_0. *X is a nonempty finite set.*

AXIOM A_1. *P is a strict partial order on X.*

THEOREM. *A_0 and A_1 imply that there is a real-valued function f on X such that, for all $x, y \in X$, xPy implies $f(x) > f(y)$.*

Although our mathematical developments need not refer to decisions and people, for interpretive purposes we shall show how these developments relate to various types of decision situations.

What are the mathematics of decision theory?

Unlike some topics such as linear algebra, general topology, and probability theory, decision-theory mathematics is not generally thought of as a well-defined special area of mathematics. People working in decision theory have a variety of backgrounds in economics, psychology, statistics, operations research, and other disciplines and they come to decision theory with varied mathematical training and interests. Consequently they have investigated many types of decision situations, using a wide variety of mathematics in these investigations. The purpose of this book is to identify and describe the main branches of mathematics that have been used and to show why and how these are used. Because of this our topics are arranged by mathematical areas rather than by typical types of decision situations.

Despite the variety of mathematics that we shall encounter, some patterns will emerge as we proceed. We shall begin with a discussion of sets since they are basic to all mathematics. Chapter 3 then discusses an immediate child of set theory that is the backbone of decision theory. This creature is the binary relation. It is used in almost all decision theories and is prominent in many areas of mathematics. For example, graph theory as typified by Berge (1958), Ore (1962), and Harary, Norman, and Cartwright (1965), is largely a study of various kinds of binary relations.

Our study of binary relations continues in Chapter 4 with the introduction of choice functions, which thus far have been used mainly in studies of consumer choice in economics and in social choice theory.

Chapter 5 is concerned with the existence of solutions for finite systems of linear inequalities and/or equalities. As we shall see in Chapter 5 the mathematics of linear systems are relevant to virtually all areas of axiomatic decision theory.

The reader who is interested only in finite-set situations will find very little to suit his taste after Chapter 5. Chapters 6 and 7 begin a rather direct confrontation with infinite sets. The former discusses Zorn's lemma and several of its equivalents. The latter involves the notion of order denseness and some basic theory of real-valued order-preserving functions.

The final four chapters of the book discuss types of structures for sets that are largely although not exclusively designed for situations involving infinite sets. These four chapters are concerned with topological spaces, probability measures, mixture sets, and ordered groupoids.

On the structure of axiomatic theories

During our excursions we shall visit many specific decision theories. To prepare for this a few general remarks are in order.

Many decision theories are based on the notion of preference. The axioms of such theories state assumptions about the structure of the set of decision alternatives and about properties of preference within this structure. Some axioms in their simplest form say nothing directly about preference, such as: X is a finite set. Some axioms that involve preference make no demands on the structure of the alternative set, such as: if x, y and z are in X and if xPy and yPz then xPz, where P means 'is preferred to'. Other axioms combine preference with structure in a variety of ways. One axiom of this type is: if xPy then there is a z in X such that xPz and zPy.

Most preference theories contain axioms that are viewed as criteria of consistent behavior or of rational introspection. Transitivity (if xPy and yPz then xPz) is one example of this. Other preference axioms may be less obvious in this respect and might be included for the explicit purpose of promoting tidiness within the axiomatic system. Many preference and non-preference structural axioms are including to impart a cohesiveness to the axiomatic system that will facilitate the derivation of foreseen theorems.

In most decision theories the axioms provide a qualitative system based on the notion of preference or of choice, and the principal theorems relate the qualitative concepts to a quantitative (numerical) system. Numerical quantities such as probabilities may appear in the axioms, but only in isolated cases are qualitative relationships absent from the axioms. Correspondingly, qualitative statements may appear as derived theorems, but only in isolated cases are quantitative relationships absent from the theorems.

The process of research in decision theory usually proceeds in one of two ways. In the 'forward' direction the researcher considers a class of decision situations having some common structure and looks for theorems that follow from various axiomatic systems built on this structure. In the 'backward' direction the researcher begins with desired representations (that specify for example a connection between qualitative concepts and a quantitative system) and tries to discover axioms that imply the existence of such representations. Obviously, both directions can be intertwined in one investigation.

If S is a proposition (representation, formula), a set of axioms is *sufficient* for S if the axioms imply S. Each axiom that is implied by S is *necessary* for S. If each axiom in a set sufficient for S is necessary for S then the set of axioms or conditions is said to be *necessary and sufficient* for S. In the backward approach we often try to find axioms that are necessary for the given S as well as sufficient for S, but often it is very difficult to do this.

To illustrate some of these ideas, consider the simple theory given earlier in this chapter. In the forward approach we could begin with axioms A_0 (X is a nonempty finite set) and A_1 (P on X is transitive, and xPx for no x in X). With sufficient patience we might then be able to prove that A_0 and A_1 imply proposition S where

S = 'there is a real-valued function f on X such that for all x and y in X, xPy implies $f(x) > f(y)$'.

We have thus proceeded from a qualitative preference system to a quantitative representation for preferences.

In the backward approach we might begin with S and find that $\{A_0, A_1\}$ is sufficient for S. However, we note also that neither A_0 nor A_1 is necessary for S. That is, S does not require X to be finite and it does not imply that P is transitive. Further examination shows that

AXIOM A_2. *There is no nonempty finite subset* $\{x_1, x_2, \ldots, x_n\}$ *of* X *for which* x_1Px_2 *and* x_2Px_3 *and* … *and* $x_{n-1}Px_n$ *and* x_nPx_1,

is implied by S. Therefore A_2 is necessary for S and it along with A_0 can be shown to imply S.

I do not presently know of a set of axioms of the type considered here that is necessary and sufficient for S. In terms defined later the difficulty arises in those instances where X is uncountable.

Methods of proof

The methods used in proving theorems in decision theory are the usual methods used throughout modern mathematics. As final preparation for what follows I shall review briefly and informally some of these methods.

The reader who is interested in a fuller discussion of the kind of formal set theory used as the foundation of our investigations and of the predicate calculus of formal logic may consult Cohen (1966).

In this discussion I shall use &, $\vee$, $\sim$, and $\Rightarrow$ to denote logical conjunction (and), disjunction (or), contradiction (not), and implication (implies). We shall say that a set of propositions is *consistent* if their logical conjunction does not imply the contradictory of one of the propositions and does not contradict some preaccepted axiom or theorem of set theory. The usual way to verify consistency is to construct a situation that we are sure is consistent and to show that each axiom in the set holds in this situation. For example, the situation where X contains a single element b and xPy for no x and y in X is logically consistent. Since both A_0 and A_1 hold in this situation $\{A_0, A_1\}$ is consistent.

If $\{A, B, \ldots\}$ is a set of propositions then A is *independent* in this set if both $\{A, B, \ldots\}$ and $\{\sim A, B, \ldots\}$ are consistent. In $\{A_0, A_1, A_2\}$ both A_0 and A_1 are independent but A_2 is not independent since it is implied by A_1.

Let $\mathcal{A}$ be a set of axioms and let S be a consistent proposition. Then $\mathcal{A} \cup \{S\}$ is the set of propositions that contains each of the axioms plus S, and $\mathcal{A} \cup \{\sim S\}$ is the set of propositions that contains each axiom plus the contradictory of S. The following four cases can arise:

1. $\mathcal{A} \cup \{S\}$ and $\mathcal{A} \cup \{\sim S\}$ are consistent.
2. $\mathcal{A} \cup \{S\}$ is consistent, $\mathcal{A} \cup \{\sim S\}$ is not consistent.
3. $\mathcal{A} \cup \{S\}$ is not consistent, $\mathcal{A} \cup \{\sim S\}$ is consistent.
4. $\mathcal{A} \cup \{S\}$ and $\mathcal{A} \cup \{\sim S\}$ are not consistent.

In the first three cases $\mathcal{A}$ is consistent and in case 4 $\mathcal{A}$ is not consistent. Henceforth I shall be concerned only with the first three cases, assuming that $\mathcal{A}$ is consistent.

For case 1 S is independent in $\mathcal{A} \cup \{S\}$. Let A denote the conjunction of the axioms in $\mathcal{A}$. Then $\mathcal{A} \cup \{S\}$ is consistent means that $\sim (A \Rightarrow \sim S)$, and $\mathcal{A} \cup \{\sim S\}$ is consistent means that $\sim (A \Rightarrow S)$. In words $\sim (A \Rightarrow S)$ reads: it is false that A implies S. As we noted above, the usual way to prove that $\sim (A \Rightarrow S)$ is to construct a situation that shows $\mathcal{A} \cup \{\sim S\}$ to be consistent. This is commonly referred to as *proof by counterexample.* To illustrate it let $A = A_0$ & A_1 and let S = 'for all x, y and z in X, if xPz then either xPy or yPz'. Let $X = \{a, b, c\}$ with $a \neq b \neq c \neq a$ and let xPy hold if and only if $x = a$ & $y = b$. Then A_0 and A_1 hold but S does not. Hence $\mathcal{A} \cup \{\sim S\}$ is consistent and $\sim (A \Rightarrow S)$. Similarly, it is easy to construct an example in which A_0 & A_1 & S holds, so that $\mathcal{A} \cup \{S\}$ is consistent and $\sim (A \Rightarrow \sim S)$.

Case 2 says that $A \Rightarrow S$ and case 3 says that $A \Rightarrow \sim S$. It is important to note that $\sim (A \Rightarrow S)$ is different than $A \Rightarrow \sim S$. That is, 'A does not imply S' is not the same as 'A implies not S'. We have just proved the first statement in the preceding example but the second is false since $\mathcal{A} \cup \{S\}$ is consistent, or $\sim (A \Rightarrow \sim S)$.

A *direct proof* of $A \Rightarrow S$ proceeds to establish S by a series of steps that begin with A or with propositions already shown to follow from A. Thus the proof may proceed through a chain of propositions, as $A \Rightarrow A' \Rightarrow A'' \Rightarrow \ldots \Rightarrow S$. A variant of this that is used frequently in decision theory may be referred to as proof by embedding or *proof by extension.* This involves embedding the structure of A in a more mathematically tractable structure. Appropriate propositions are verified within the new structure and S is shown to follow from these and the constructions used in the extension. Chapter 3 shows how the theorem at the beginning of this chapter is proved by extension.

A *proof by contradiction* (indirect proof, reductio ad absurdum) for $A \Rightarrow S$ consists of showing that $\mathcal{A} \cup \{\sim S\}$ is inconsistent, which means that $\sim S \Rightarrow \sim A$ (which is logically equivalent to $A \Rightarrow S$). Thus we suppose that S is false and show that some axiom in A must be false. To illustrate let $A = A_0$ & A_1 and let S = 'for all x and y in X, if xPy then not yPx'. Contrary to S suppose that there are a and b in X such that aPb and bPa. Then by the transitivity part of A_1 we have aPa, which contradicts the other part of A_1. Hence A_0 & $A_1 \Rightarrow S$.

We now describe some special cases that may involve either direct or indirect proofs, or both. Suppose first that S is itself the conjunction of a number of other propositions. A proof that proves each of these separately is one form of a *proof by exhaustion* or a *proof by enumeration.* In symbols: $S = S_1$ & S_2 & ... & S_n, and if $(A \Rightarrow S_1)$ & $(A \Rightarrow S_2)$ & ... & $(A \Rightarrow S_n)$ then $A \Rightarrow S$.

In some composite S cases our task may be greatly simplified by using the *principle of induction.* Thus suppose that $S = S_1$ & S_2 & ... & S_n or else that $S = S_1$ & S_2 & ... (with no end). Suppose we can prove that $A \Rightarrow S_1$ and that whenever S_i and S_{i+1} are in the sequence $S_1, S_2, \ldots$ then A & $S_i \Rightarrow$ $\Rightarrow S_{i+1}$. The principle of induction then says that $A \Rightarrow S$. For example, suppose we wish to prove that the axioms of set theory and the usual properties of integers that follow therefrom imply S = 'if n is a positive integer then $1+2+\ldots+n = n(n+1)/2$'. Let S_n be the proposition '$1+2+\ldots+n = n(n+1)/2$' so that $S = S_1 \& S_2 \& \ldots$. S_1 is obviously true. Suppose then that S_n is true so that $1+2+\ldots+n = n(n+1)/2$. Then $(1+2+\ldots+n) +(n+1) = n(n+1)/2+(n+1) = (n+1)(n+2)/2$ so that S_{n+1} is true also.

Corresponding methods apply when A can be written as the disjunction of a number of propositions. Suppose in the above example that along with the axioms of set theory we include explicitly the proposition A = 'n is a positive integer' so that $A = A_1 \vee A_2 \vee \ldots$ where A_k = '$n = k$' and take S = '$1+2+\ldots+n = n(n+1)/2$'. We then show that $A_1 \Rightarrow S$ and that if $A_k \Rightarrow S$ then $A_{k+1} \Rightarrow S$. We then have $A_1 \Rightarrow S$ and $(A_k \Rightarrow S) \Rightarrow (A_{k+1} \Rightarrow S)$ for $k = 1, 2, \ldots$, and therefore $A \Rightarrow S$.

In many cases a theorem $A \Rightarrow S$ is stated in which a set X appears in A but its size is not specifically prescribed by A or by S. The method of disjunctive exhaustion might then be used by writing, for example, A = (A & X is empty) $\vee$ (A & X is nonempty and finite) $\vee$ (A & X is infinite), and proving that each (A & X is ...) implies S.

CHAPTER 2

sets and functions

This chapter defines and illustrates some basic notions of set theory that are used in decision theory. The next chapter continues with a discussion of binary relations.

Sets are composed of elements with the exception of the *empty set* ϕ that contains no element. $x \in X$ means that x is an element in X, or that x is a member of X. $x \notin X$ means that x is not in X. When $x \in X$ and $y \in X$, we shall often write $x, y \in X$.

Braces are often used to identify sets. Thus $\{1, 2, 3, \ldots\}$ is the set of *positive integers*, and $\{\phi\}$ is the set whose only element is the empty set. $\{x : C\}$ denotes the set of all x that satisfy condition C. Thus $\{x : x$ is a man & x is alive & x is married$\}$ is simply the set of living married men. $\phi = \{x : x$ is a real number & $x < 0$ & $x > 0\}$.

The set A is a *subset* of B if every element in A is in B. $A \subseteq B$ means that A is a subset of B, or that A is included in B, or that B includes A. If $x \in X$ then $\{x\} \subseteq X$. Since ϕ has no elements, every element in ϕ is in any other set. Hence $\phi \subseteq X$ and $\phi \subseteq \phi$.

Two sets A and B are identical if and only if $A \subseteq B$ and $B \subseteq A$, in which case $A = B$. A is a *proper subset* of B, written $A \subset B$, if and only if $A \subseteq B$ and $B \nsubseteq A$. Thus $\{1, 2, 3, \ldots\} \subset \{0, 1, 2, 3, \ldots\}$. In most elections the set of voters is a proper subset of the set of eligible voters. $\{$Abraham Lincoln, Napoleon Bonaparte$\} \subset \{x : x$ is a deceased head of state$\}$.

If $A \subseteq B$ and $B \subseteq C$ then $A \subseteq C$. Thus $\subseteq$ is a basic example of a transitive relation. We shall study these further in the next chapter.

Unions, intersections and partitions

$A \cup B$, the *union* of A and B, is the set of all elements in A or in B. Thus $A \cup B = \{x : x \in A \text{ or } x \in B\}$. In all cases $A = A \cup A$, $A \subseteq A \cup B$, and $A \cup B = B \cup A$. $A = B$ if and only if $A \cup B \subseteq A$ and $A \cup B \subseteq B$, since in general $C \subseteq D$ if and only if $C \cup D \subseteq D$. $\{-1, -2, -3, \ldots\} \cup \{1, 2, 3, \ldots\}$ is the set of all nonzero integers. $A \cup B = \phi$ if and only if $A = \phi$ and $B = \phi$.

The union of many sets is the set of all elements in at least one of the sets. When $\mathcal{A}$ is a set of sets, their union may be written as $\bigcup_{\mathcal{A}} A$. Thus if $\mathcal{A} = \{\{1\}, \{2\}, \{3\}, \ldots\}$ then $\bigcup_{\mathcal{A}} A = \{1, 2, 3, \ldots\}$. Another notation is often used. If $A_1, A_2, \ldots, A_n$ are sets, their union may be written as $A_1 \cup A_2 \cup \ldots \cup A_n$ or as $\bigcup_{i=1}^{n} A_i$. More generally, if I is a set and $A(i)$ is a set for each $i \in I$ then the union of these $A(i)$ may be written as $\bigcup_I A(i)$.

$A \cap B$, the *intersection* of A and B, is the set of all elements in *both* A and B. Thus $A \cap B = \{x : x \in A \,\&\, x \in B\}$. In all cases $A = A \cap A$, $A \cap B \subseteq A$, and $A \cap B = B \cap A$. $A \subseteq B$ if and only if $A \cap B = A$. If A is the set of all women and B is the set of all married people then $A \cap B$ is the set of all married women.

The intersection of many sets is the set of elements in every one of the sets. $\bigcap_{\mathcal{A}} A$, $A_1 \cap A_2 \cap A_3 \cap \ldots$, $\bigcap_{i=1}^{n} A_i$, and $\bigcap_I A(i)$ are notations for intersections defined like those for unions.

As you can readily verify, $\cup$ and $\cap$ satisfy the two distributive laws

$$(A \cup B) \cap C = (A \cap C) \cup (B \cap C)$$

and

$$(A \cap B) \cup C = (A \cup C) \cap (B \cup C).$$

Sets A and B are *disjoint* (have no common element) if and only if $A \cap B = \phi$. The sets in a set $\mathcal{A}$ of sets are *mutually disjoint* if $A \cap B = \phi$ whenever $A, B \in \mathcal{A}$ and $A \neq B$. Thus the sets in $\mathcal{A} = \{\{1\}, \{2\}, \{3, 4, 5\}\}$ are mutually disjoint.

A *partition* of a set A is a set of nonempty and mutually disjoint subsets of A whose union equals A. There is no partition of ϕ. If $A \neq \phi$, $\{A\}$ is a partition of A. $\{\{1, 3, 5, \ldots\}, \{2, 4, 6, \ldots\}\}$ is a two-part partition of the set of positive integers. An *n-part partition* of a set is a partition with n elements (subsets).

Complements and differences

When $A \subseteq X$, the *complement* of A in X is

$$A^c = \{x : x \in X \;\&\; x \notin A\}.$$

If $\phi \subset A \subset X$ then $\{A, A^c\}$ is a two-part partition of X.

If A and B are any two sets, the directed *difference* $A-B$ is

$$A-B = \{x : x \in A \;\&\; x \notin B\}.$$

If A and B are subsets of X then $A-B = A \cap B^c$ and

$$(A \cup B)^c = A^c \cap B^c \text{ and } (A \cap B)^c = A^c \cup B^c,$$

so that the complement of a union is the intersection of the complements and the complement of an intersection is the union of the complements. These two formulas were independently discovered by Augustus De Morgan and Benjamin Pierce in the last century and are known today as De Morgan's laws. The second law follows from the first and the observation that $A^{cc}=A$: by the first law $(A^c \cup B^c)^c = A \cap B$, so that $A^c \cup B^c = (A \cap B)^c$, which is the second law.

Products of sets

The *product* (Cartesian product) $A \times B$ of sets A and B is the set of all ordered pairs (a, b) with $a \in A$ and $b \in B$. The *ordered* pair (a, b) with *first component* a and *second component* b is defined in set terms by $\{\{a\}, \{a, b\}\}$, which is the same as $\{\{a, b\}, \{a\}\}$. This brings out the fact that $(a, b) \neq (b, a)$ unless $a = b$, in which case $\{\{a\}, \{a, b\}\} = \{\{a\}, \{a, a\}\} = \{\{a\}, \{a\}\} = \{\{a\}\}$. $A \times B = \{(a, b) : a \in A \;\&\; b \in B\}$.

$\{1, 3, 5, \ldots\} \times \{2, 4, 6, \ldots\}$ is the set of all ordered pairs of positive integers with odd first component and even second component. The set of all *rational numbers* (ratios of integers) can be defined in terms of ordered pairs as follows. Let $C = \{\ldots, -2, -1, 0, 1, 2, \ldots\} \times \{\ldots, -2, -1, 1, 2, \ldots\}$. We associate $(p, q) \in C$ with the ratio p/q. We now partition C by letting (p, q) and (r, s) be in the same element of the partition if and only if $ps = rq$. Each element of this partition is, by definition, a rational number. Thus

$\{\{1, 2\}, \{-1, -2\}, \{2, 4\}, \{-2, -4\}, \ldots\}$ is a rational number: it is commonly represented by any one of its elements, such as (1, 2), or 1/2. For brevity we refer to 1/2 as a rational number.

The product of sets $A_1, A_2, \ldots, A_n$ is the set $\{(a_1, a_2, \ldots, a_n) : a_i \in A_i$ for $i = 1, 2, \ldots, n\}$ of *n-tuples* $(a_1, a_2, \ldots, a_n)$. This product is written as $A_1 \times A_2 \times \ldots \times A_n$, or $\Pi_{i=1}^n A_i$, or ΠA_i. The product of set A with itself n times is $A^n = \{(a_1, \ldots, a_n) : a_i \in A$ for $i = 1, \ldots, n\}$.

Products of sets are very common in decision theory. They arise in static multiple-factor analyses (each A_i is the set of levels of a particular factor or performance characteristic of a system), in dynamic theories (A_i is the set of outcomes in the ith time period), in social choice theory (A_i is the set of possible values of the ith voter), in game theory (A_i is the set of possible strategies for player i), and in other contexts.

Real numbers

The set of real numbers, denoted here as *Re*, consists of the rational numbers and the irrational numbers (those that are not rational). Thus, the rationals and irrationals form a two-part partition of *Re*. $\sqrt{2}$ is irrational, for suppose to the contrary that $\sqrt{2} = p/q$, where p and q are positive integers with no common factor: there are no positive integers a, b, c with $a > 1$, $ab = p$, and $ac = q$. Then $2q^2 = p^2$, which implies that p is an even integer, say $p = 2r$, and hence q must be odd. But then $q^2 = 2r^2$ which requires q to be even, and we arrive at a contradiction.

Without going into much detail we note that real numbers are defined in terms of subsets of rational numbers. For example, a real number can be defined as a subset A of rational numbers that (1) is nonempty, (2) is bounded above ($a < b$ for all $a \in A$ and some rational b), (3) has no largest element (if $a \leq b$ for all $a \in A$ and b is rational then $b \notin A$), and (4) if x is rational and $x < a$ for some $a \in A$ then $x \in A$. Thus 1/2 is characterized by $\{x : x$ is rational and $x < 1/2\}$, and $\sqrt{2}$ by $\{x : x$ is a nonpositive rational$\} \cup \{x : x$ is a positive rational and $x^2 < 2\}$. In such terms A is "rational" if $B = \{x : x$ is rational and $a < x$ for all $a \in A\}$ has a smallest element, in which case A is represented by this element. If B has no smallest element then A is 'irrational'.

Under the usual axiomatic development of the real number system we find the following useful properties: if x is a real number and $A = \{a : a \in Re$

& $a < x\}$ and $B = \{b : b \in Re$ and $b > x\}$, then $\{A, B, \{x\}\}$ is a partition of *Re*; if $x, y \in Re$ and $x < y$ then there is a rational number r such that $x < r < y$; if $a, b \in Re$ and $a > 0$ and $b > 0$ then $na > b$ for some positive integer n. The last of these is called the *Archimedean law*. It has a number of counterparts in decision theory axiomatics as we shall see later.

A nonempty subset B of *Re* is *bounded below (above)* if there is an $x \in Re$ such that $x \leq b$ $(b \leq x)$ for all $b \in B$, in which case x is a *lower (upper) bound* on B. If $B \neq \phi$ is bounded below, its greatest lower bound or *infimum* is the unique real number inf B that is a lower bound on B and is as large as every other lower bound on B. If $B \neq \phi$ is bounded above, its least upper bound or *supremum* is the unique real number sup B that is an upper bound on B and is not greater than any other upper bound on B. Thus, if $B = \{x : 0 < x < 1\}$ then inf $B = 0$ and sup $B = 1$. If $B \neq \phi$ and B has no lower (upper) bound, this is often expressed by writing inf $B = -\infty$ (sup $B = \infty$).

An *interval* of real numbers is a nonempty subset $A \subseteq Re$ having the property that if $a < x < b$ and $a, b \in A$ then $x \in A$. With $a < b$ we write

$[a, b] = \{x : a \leq x \leq b\}$, the closed interval with end points a and b;
$]a, b[= \{x : a < x < b\}$, the open interval with end points a and b;
$[a, b[= \{x : a \leq x < b\}$, a half-open interval;
$]a, b] = \{x : a < x \leq b\}$, another half-open interval.

In addition, $[a, \infty[= \{x : a \leq x\}$; $]a, \infty[= \{x : a < x\}$; $]-\infty, b] = \{x : x \leq b\}$; $]-\infty, b[= \{x : x < b\}$; $]-\infty, \infty[= Re$.

Re^n, known as *n-dimensional Euclidean space* or Euclidean n-space, is defined as indicated above: $Re^n = \{(x_1, \ldots, x_n) : x_i \in Re$ for $i = 1, \ldots, n\}$. The n-tuples in Re^n are also called n-dimensional *vectors*. If $x = (x_1, \ldots, x_n)$ and $y = (y_1, \ldots, y_n)$ are in Re^n and if $\alpha, \beta \in Re$ then multiplication of vectors by scalars (real numbers) and vector addition are defined by

$$\alpha x + \beta y = (\alpha x_1, \ldots, \alpha x_n) + (\beta y_1, \ldots, \beta y_n) = (\alpha x_1 + \beta y_1, \ldots, \alpha x_n + \beta y_n),$$

so that $\alpha x + \beta y \in Re^n$. Moreover, the *inner product* (dot product, scalar product) of x and y is defined by

$$x \cdot y = x_1 y_1 + \ldots + x_n y_n = \textstyle\sum_{i=1}^{n} x_i y_i,$$

so that $x \cdot y \in Re$.

$x \in Re^n$ is *rational* if each of its components is a rational number. $x \in Re^n$ is *integral* if each of its components is an integer.

Functions

A function f on a set X *to* a set Y is a subset of the product $X \times Y$ such that if $x \in X$ then $(x, y) \in f$ for one and only one $y \in Y$. (What I am calling a function is sometimes referred to as a single-valued function.)

When f is a function on X to Y, we write $f(x) = y$ to mean the same thing as $(x, y) \in f$. If the set of all second components of the ordered pairs in f exhausts Y, so that $\{f(x) : x \in X\} = Y$, then f is a function on X *onto* Y. f on X to Y has an inverse $f^{-1} = \{(y, x) : (x, y) \in f\}$ if and only if (x, y), $(x', y) \in f$ implies $x = x'$, in which case f^{-1} is on $\{f(x) : x \in X\}$ onto X. If f is on X *onto* Y and f has an inverse, then f is one to one, or a *one to one correspondence between X and Y*. This says that f assigns a unique $y \in Y$ to each $x \in X$ and for each $y \in Y$ there is a unique $x \in X$ for which $f(x) = y$. A one to one correspondence between X and X is a *permutation* of X. Thus, if the x_i are all distinct ($i = 1, \ldots, n$), then $\{(x_1, x_2), (x_2, x_3), \ldots, (x_{n-1}, x_n), (x_n, x_1)\}$ is a permutation of $\{x_1, x_2, \ldots, x_n\}$. f is a permutation of $\{a, b, c\}$ when $f(a) = a$, $f(b) = c$, and $f(c) = b$.

A function f on X to Re is a *real-valued function*, and in this case we say that f is a real-valued function on X. A real-valued function f on X assigns a real number $f(x)$ to each $x \in X$. Such an f is *bounded below (above)* if $\{f(x) : x \in X\}$ has a lower (upper) bound, and it is *bounded* if it is bounded both below and above.

If f and g are real-valued functions on X then g is an *order-preserving transformation* of f if and only if, for any $a, b \in X$, $f(a) < f(b)$ implies $g(a) < g(b)$ and conversely. g is a *linear transformation* of f if there are real numbers $\alpha \neq 0$ and β such that $g(x) = \alpha f(x) + \beta$ for all $x \in X$. We shall call an order-preserving linear transformation a *positive* ($\alpha > 0$) *linear transformation.*

Finite and infinite sets

A set X is *finite* if and only if it is empty or else there is a positive integer n such that there is a one to one correspondence between X and $\{1, 2, \ldots, n\}$. When there is a one to one correspondence between X and $\{1, 2, \ldots, n\}$, n is the cardinality or power of X, which simply means that X has n elements.

A set X is *denumerable* if and only if there is a one to one correspondence between X and the set $\{1, 2, 3, \ldots\}$ of all positive integers. Examples of denumerable sets are: the set of all integers; the set of rational numbers; the set $A \times B$ when either A or B is denumerable and the other is not empty and is either finite or denumerable.

X is *countable* if and only if it is finite or denumerable. By far the most famous example of a set that is not countable is *Re*, the set of real numbers. A set that is not countable is *uncountable*. If $A \subseteq B$ and A is uncountable then so is B. Suppose for example that the set $\{.a_1a_2a_3\ldots : a_i \in \{0, 1\}$ for $i = 1, 2, 3, \ldots\}$ of decimals whose terms are zeros and ones is countable, and let the necessary one to one correspondence between this set and $\{1, 2, 3, \ldots\}$ be

$$(1, .a_{11}a_{21}a_{31}\ldots)$$
$$(2, .a_{12}a_{22}a_{32}\ldots)$$
$$(3, .a_{13}a_{23}a_{33}\ldots)$$
etc.

Then let $b_i = 1 - a_{ii}$ for $i = 1, 2, \ldots$. Then $.b_1b_2b_3\ldots$ is in our original set but is different than each of the decimals in the correspondence, which contradicts the existence of such a correspondence. Hence the original set of decimals is uncountable. Since *Re* includes this set it too is uncountable.

As we shall see, axioms of a decision theory may differ according to whether basic sets used in the axioms are finite, denumerable, or uncountable.

CHAPTER 3

binary relations

Virtually all axiomatic decision theories use binary relations. This chapter identifies some of the types of binary relations that arise in these theories.

Basic definitions

A *binary relation* R on a set X is a subset of ordered pairs in the product $X \times X$. Thus R is a binary relation on X if and only if $R \subseteq X \times X$. ϕ denotes the *empty relation* and $\Delta = \{(x, x) : x \in X\}$ is the *equality relation.* $(x, y) \in \Delta$ if and only if $x = y$.

We write xRy to mean the same thing as $(x, y) \in R$ and say that x has the relation R to y. We shall use P to denote preference: xPy if and only if x is preferred to y. If X is a set of legislators we might interpret xRy as x is more powerful than y.

$(x, y) \notin R$ is also written as not xRy. If R is a binary relation on X then either $(x, y) \in R$ or $(x, y) \notin R$, and not both. Consequently, exactly one of the following holds when $x, y \in X$:

1. xRy & yRx
2. xRy & not yRx
3. not xRy & yRx
4. not xRy & not yRx.

Either 1 or 4 holds when $x = y$. If X is a set of people and R means 'is taller than' then 1 is impossible and 4 says that x and y are of equal height. Each of the four possibilities can arise when xRy means that x likes y. 2 and 3 denote essentially the same thing since 3 results from the interchange of x and y in 2.

The *dual* or *converse* of R is defined as

$$R^* = \{(x, y) : (y, x) \in R\},$$

and the *complement* of R is

$$R^c = \{(x, y) : (x, y) \in X \times X \quad \& \quad (x, y) \notin R\}.$$

An important property of duals and complements is that $R^{c*} = R^{*c}$.

$\Delta^* = \Delta$ and $\Delta^c = \{(x, y) : x \neq y\}$. If R means 'is taller than' then R^* means 'is shorter than' and R^c means 'is not taller than'. Or with $<$ the natural order on Re, $<^* = >$ and $<^c = \geqslant$.

The *composition* (relative product) of two binary relations R and S on X is the binary relation RS defined by

$$RS = \{(x, y) : (x, a) \in R \quad \& \quad (a, y) \in S \quad \text{for some} \quad a \in X\}.$$

Thus $\Delta\Delta = \Delta$, the composition of $<$ and $<$ is $<$, and the composition of $<$ and $>$ is Re. In many cases RS is different than the intersection $R \cap S = \{(x, y) : (x, y) \in R \text{ and } (x, y) \in S\} : < \cap > = \phi$.

Some relational properties

A binary relation R on X may or may not have certain special properties. Some of the special properties that have been used in decision theories are grouped into five categories as follows. In each case except P10 the property is expressed in two equivalent ways. The second expressions are to hold for all $x \in X$, or all $x, y \in X$, and so forth.

P1. *reflexivity*. $\Delta \subseteq R$. xRx.
P2. *irreflexivity*. $R \subseteq \Delta^c$. not xRx.
P3. *symmetry*. $R^* = R$. xRy implies yRx.
P4. *asymmetry*. $R \cap R^* = \phi$. xRy implies not yRx.
P5. *antisymmetry*. $R \cap R^* \subseteq \Delta$. xRy & yRx imply $x = y$.
P6. *transitivity*. $RR \subseteq R$. xRy & yRz imply xRz.
P7. *negative transitivity*. $R^cR^c \subseteq R^c$. not xRy & not yRz imply not xRz. Equivalently: xRz implies xRy or yRz.

P8. *connectedness* (completeness). $R \cup R^* = X \times X$. xRy or yRx.
P9. *weak connectedness*. $\Delta^c \subseteq R \cup R^*$. $x \neq y$ implies xRy or yRx.
P10. xRy & zRw imply xRw or zRy.
P11. $RR \subseteq (R^cR^c)^c$. xRy & yRz imply xRw or wRz.

The final two properties come from Scott and Suppes (1958) based on prior work by Luce (1956). Chipman's (1971) terminology has been used for P7.

An increased familiarity with P1 through P11 can be gained by proving the following facts: the two characterizations in P7 are equivalent; R is irreflexive if it is asymmetric; R is asymmetric if it is irreflexive and transitive; R is transitive if it is asymmetric and negatively transitive; R is transitive if it is irreflexive and satisfies P10 or P11; R is antisymmetric if it is asymmetric; R is negatively transitive if it is irreflexive, transitive and weakly connected; R is connected if it is reflexive and weakly connected; R satisfies P10 and P11 if it is asymmetric and negatively transitive.

'Is taller than' has properties P2, P4, P5, P6, P7, P10, and P11. $<$ on *Re* has the same properties plus P9. $=$ on *Re* has properties P1, P3, P5, and P6. $\leq$ on *Re* has properties P1, P5, P6, P7, P8, P9, P10, and P11. $\subset$ (proper inclusion) on the set of all subsets of a set X has properties P2, P4, P5, and P6. 'Likes' may have none of the listed properties.

Some compatible combinations of the foregoing properties have special names because of their widespread use in mathematics. We shall look at such combinations in the following pages. When I name a combination, I shall note in parentheses some other names that have been used for the same combination.

Equivalence relations

An *equivalence* (equivalence relation) E on a set X is a binary relation on X that is reflexive (xEx), symmetric (if xEy then yEx) and transitive (if xEy & yEz then xEz). $=$ on *Re* is an equivalence.

Any equivalence E on X determines a partition of X in the following way: x and y are in the same element of the partition if and only if xEy. With $E(x) = \{y : xEy\}$, this partition is

$$X/E = \{E(x) : x \in X\}.$$

For any $E(x)$ and $E(y)$ either $E(x) \cap E(y) = \phi$, if not xEy, or $E(x) = E(y)$

if xEy. X/E is the set of *equivalence classes* of X under E, and $E(x)$ is the *equivalence class generated by* x.

> $(Re/=) = \{\{x\} : x \in Re\}$. If xEy if and only if $x, y \in X$ then $X/E = \{X\}$.

If xEy means that $x, y \in\ ]k, k+1]$ for some integer k then $X/E = \{]k, k+1] : k \in \{\ldots, -2, -1, 0, 1, 2, \ldots\}\}$.

Conversely, any partition H of X determines an equivalence E on X as follows:

$$E = \{(x, y) : x, y \in a \quad \text{for some} \quad a \in H\},$$

with $X/E = H$. If H is the partition of a set of registered voters according to the parties they can register under then E as here defined has xEy if and only if x and y are registered in the same party.

The intersection of an arbitrary number of equivalences on a given set X is an equivalence on X. Thus if X is a set of social alternatives, if $(x, y) \in I_i$ if and only if voter i is indifferent between x and y, and if each I_i is an equivalence, then the set $\cap I_i$ of all unanimous indifference pairs is an equivalence.

Let I on a set X of mutually exclusive decision alternatives be interpreted as xIy if and only if the decision maker is *indifferent* between x and y. Many decision theories assume that I on X is an equivalence, and in this case the elements in X/I are *indifference classes.*

Armstrong (1939) challenges the assumption of transitive indifference. For example, a person may be indifferent between 100 and 101 grains of sugar in his coffee, indifferent between 101 and 102, ..., and indifferent between 4999 and 5000. If indifference were transitive he would be indifferent between 100 and 5000 grains, and this is probably false.

Even when indifference is not assumed to be transitive it may be possible to define an equivalence on X based on I. Suppose P (preference) on X is asymmetric and I denotes the absence of P so that $I = (P \cup P^*)^c = P^c \cap P^{*c} = P^c \cap P^{c*}$. That is, xIy if and only if not xPy and not yPx. Then E, with xEy if and only if (xIz if and only if yIz, for all $z \in X$), is an equivalence on X. In this case

$$E = (II^c \cup I^c I)^c.$$

Order relations

The phrase 'order relation' usually refers to a transitive binary relation. Thus, an equivalence is an 'order relation' although it 'orders' a set only in the sense of sorting its elements into equivalence classes.

The order relations discussed in this chapter conveniently dichotomize themselves into those that are asymmetric (and hence irreflexive) and those that are reflexive (and hence not asymmetric). The next section discusses asymmetric orders, after which reflexive orders will be considered.

Asymmetric orders

To display a range of asymmetric orders used in decision theory, we shall examine a hierarchy of six of these. The first is actually not an 'order' since it is not assumed to be transitive. For its definition we define the *transitive closure* of a binary relation P by $P^t = P \cup PP \cup PPP \cup \ldots$ so that xP^ty if and only if xPy or there are $x_1, \ldots, x_m$ such that xPx_1 & x_1Px_2 & ... & $x_{m-1}Px_m$ & x_mPy. In the following definitions we may think of P as 'is preferred to' or 'is more probable than' or 'is more powerful than', and so forth.

01. *suborder:* P^t is irreflexive.
02. *strict partial order* (partial order): P is irreflexive and transitive.
03. *interval order:* P is irreflexive and satisfies P10.
04. *semiorder;* P is irreflexive and satisfies P10 and P11.
05. *weak order:* P is asymmetric and negatively transitive.
06. *linear order* (strict, strong, simple, total, or complete order; chain): P is irreflexive, transitive, and weakly connected.

When P is a suborder, P^t is a strict partial order. Early uses of suborders were made in game theory by von Neumann and Morgenstern (1947, pp. 590–609), who used the term 'acyclic' for the property that P^t is irreflexive, and by Houthakker (1950, 1961) in his 'strong axiom of revealed preference'. A quantitative correspondent of the semiorder concept was used by Armstrong (1939) in his study of intransitive indifference and later received a rigorous axiomatic treatment by Luce (1956) [who introduced the term 'semiorder'] and Scott and Suppes (1958).

The prime example of a strict partial order that is not also an 03, 04, 05, or 06 order is the relation $\subset$ on the set of all subsets of a set containing more than three elements. The prime examples of linear orders are $>$ on *Re* and its converse $<$.

Weak orders have been used extensively in all areas of decision theory. Strict partial orders have received increasing attention in the past few years. P is a weak order if and only if it is a strict partial order and $I = (P \cup P^*)^c$ is transitive. When P is a weak order, P' defined on P/I by $aP'b$ if and only if xPy for some $x \in a$ and $y \in b$ is a linear order.

To demonstrate the hierarchy of these orders let X be a set and let $\{0k\}$ denote the set of all relations on X that have the properties specified by $0k$. Then it is easily verified that

$$\{06\} \subseteq \{05\} \subseteq \{04\} \subseteq \{03\} \subseteq \{02\} \subseteq \{01\}.$$

Thus, every linear order is a weak order, every weak order is a semiorder, and so forth. Conversely, let $\{0j\} \to \{0k\}$ mean that whenever $P \in \{0j\}$ there is a $P' \in \{0k\}$ such that $P \subseteq P'$, or xPy implies $xP'y$. Such a P' is an *extension* of P and P is said to be *extended to* or *embedded in* P'. It is then true that

$$\{01\} \to \{02\} \to \{03\} \to \{04\} \to \{05\} \to \{06\}.$$

The proof of this rests on the facts that $P^t \in \{02\}$ when $P \in \{01\}$ and that every $P \in \{02\}$ has an extension in $\{06\}$. The latter fact is proved by Szpilrajn (1930). To see, for example, that $\{03\} \to \{04\}$, we note that $P \in \{03\}$ is also in $\{02\}$ and hence has an extension $P' \in \{06\}$ which is also in $\{04\}$. Szpilrajn's extension theorem uses Zorn's lemma which we shall examine later.

The hierarchical features can be illustrated in two more ways. In the following list we note properties that hold for every $P \in \{0k\}$ but that do not in general hold for every $P \in \{0(k-1)\}$. $I = (P \cup P^*)^c$, and xEy means that (xIz if and only if yIz, for all $z \in X$).

01. E is an equivalence.
02. $PE \subseteq P$ and $EP \subseteq P$.
03. PI and IP are weak orders.
04. $PI \cup IP$ is a weak order.
05. I is an equivalence, $PI \subseteq P$ and $IP \subseteq P$.
06. $I = \Delta$.

For 02, suppose $(x, z) \in PE$, so that $(x, y) \in P$ & $(y, z) \in E$ for some $y \in X$, and suppose that, contrary to the conclusion, $(x, z) \notin P$. Then either xIz, in which case yIx by $(y, z) \in E$, and this contradicts $(x, y) \in P$; or zPx and hence zPy by transitivity, which contradicts $(y, z) \in E$. Hence $PE \subseteq P$ if P is a strict partial order.

Our second list assumes that X is finite and, for each $0k$, states a real-valued representation theorem that is implied by the conditions of $0k$ and which except for 02 implies the conditions of $0k$. Thus, for example, the theorem for 01 says that there is a real-valued function f on X such that, for all $x, y \in X$, xPy implies $f(x) > f(y)$, if and only if P on X is a suborder.

01. xPy implies $f(x) > f(y)$.
02. xPy implies $f(x) > f(y)$, and xEy implies $f(x) = f(y)$.
03. xPy if and only if $f(x) > f(y)+\sigma(y)$, and $\sigma(x) > 0$ for all x.
04. xPy if and only if $f(x) > f(y)+1$.
05. xPy if and only if $f(x) > f(y)$.
06. xPy if and only if $f(x) > f(y)$, and $f(x) = f(y)$ implies $x = y$.

Theorem 04 was first proved by Scott and Suppes (1958). Scott (1964) gives an alternate proof. Theorem 03 is proved in Fishburn (1970) which contains proofs of the other theorems.

To prove Theorem 06 with $X \neq \phi$ and finite, we note first that X must contain an element x_1 such that xPx_1 for all $x \neq x_1$ in X. Then $X-\{x_1\}$, if it is not empty, must contain an x_2 such that xPx_2 for all $x \in X-\{x_1, x_2\}$. Continuing this process we eventually exhaust X and get say $x_nPx_{n-1}P\ldots Px_2Px_1$. Then f with $f(x_k) = k$ satisfies the representation for 06.

Theorems 02 and 05 then follow from Theorem 06. For 02, the properties that E is an equivalence and that $PE \subseteq P$ and $EP \subseteq P$ lead to the conclusion that P' on X/E, defined by $aP'b$ if and only if xPy for some $x \in a$ and $y \in b$, is a strict partial order and that if $x \in a$ & $y \in b$ & xPy then $aP'b$. By Szpilrajn's extension theorem P' on X/E has a linear order extension P'' on X/E. Hence, by Theorem 06, there is a real-valued function g on X/E such that $aP''b$ if and only if $g(a) > g(b)$. Let $f(x) = g(a)$ whenever $x \in a$. It follows that $f(x) = f(y)$ if and only if xEy, and that with $x \in a$ and $y \in b$, xPy implies $aP'b$ implies $aP''b$ implies $g(a) > g(b)$ implies $f(x) > f(y)$.

This proof is representative of many proofs for suborders and strict partial orders. The tactic, as illustrated, is to embed the given structure in a more restrictive structure (such as a linearly ordered set) and then to use a theorem for the more restrictive structure to obtain the desired result for the original structure.

In fact, the embedding device is very common throughout mathematics and will appear in various forms in later parts of this book.

One final point on the foregoing representation theorems is in order. All these theorems, except for 04, remain valid when X is assumed to be countable and not necessarily finite. To see why 04 can fail in this case suppose P is $>$ on the set of all rational numbers. Then P is a semiorder since it is a linear order. Suppose the representation xPy if and only if $f(x) > f(y)+1$ holds, and consider $f(0)$ and $f(1)$. Since for any positive integer n there is a subset of n rational numbers between 0 and 1 we must then have $f(1) > f(0) + n$ for $n = 1, 2, \ldots$, and this is clearly impossible if $f(1)$ and $f(0)$ are numbers.

Reflexive orders

With (P, I) interpreted as (preference, indifference) we define a new relation $R = P \cup I$, with xRy if and only if xPy or xIy, and call R a preference-or-indifference relation. If I is reflexive then R must be reflexive. Moreover, if P is asymmetric and $I = (P \cup P^*)^c$ then exactly one of $(x, y) \in P$, $(y, x) \in P$, and $(x, y) \in I$ holds for any $\{x, y\} \subseteq X$ and R is complete or connected (P8).

Many axiomatic decision theories, instead of taking P as basic and defining I and R from P as indicated above, take a reflexive order relation R as the basic relation and define P and I from R as follows:

$P = R \cap R^{c*}$: xPy if and only if xRy & not yRx,
$I = R \cap R^*$: xIy if and only if xRy & yRx.

The four most common combinations of properties used for R in this approach are:

07. *preorder* (quasi order, partial order): R is reflexive and transitive.
08. *complete preorder* (weak order, simple order): R is transitive and complete.
09. *partial order* (order): R is reflexive, transitive, and antisymmetric.
010. *complete order* (simple order, chain, total order): R is transitive, complete (connected), and antisymmetric.

The prime example of a partial order is $\subseteq$ and the prime example of a complete order is $\leqslant$. The partial hierarchy in this case is $\{010\} \subseteq \{08\} \subseteq \{07\}$

and $\{010\} \subseteq \{09\} \subseteq \{07\}$. A complete preorder (08) is not necessarily a partial order (09), and conversely.

Two of these four orders have direct correspondents in the former list. With P and I defined from R as indicated, P is a weak order (05) if R is a complete preorder (08), and with $R = P \cup (P \cup P^*)^c = P \cup P^{*c}$, R is a complete preorder if P is a weak order. P is a linear order (06) when R is a complete order (010) and $P = R \cap R^{c*}$, and R is a complete order when P is a linear order and $R = P \cup P^{c*}$.

When R is either a preorder (07) or a partial order (09) and when $P = R \cap R^{c*}$ and $I = R \cap R^*$, then

(i) there may be $x, y \in X$ such that no one of xPy, yPx, and xIy holds, in which case x and y are *incomparable;*
(ii) I is an equivalence, and $I = \Delta$ for 09;
(iii) P is a strict partial order;
(iv) $PI \subseteq P$ and $IP \subseteq P$.

Despite the fact that $I = R \cap R^*$ is transitive when R is a preorder, there is a correspondence between preorders and strict partial orders. In particular, if P is a strict partial order and if E is defined as in the preceding section then, with $R = P \cup E$ (instead of $P \cup I$), R is a preorder, $(x, y) \in R \cap R^*$ if and only if $(x, y) \in E$, and $(x, y) \in P$ if and only if xRy & not yRx. Since E is transitive in this case and $EP \cup PE \subseteq P$, the transitivity of $R = P \cup E$ follows easily:

$$RR = (P \cup E)(P \cup E) = PP \cup EP \cup PE \cup EE \subseteq P \cup P \cup P \cup E$$
$$= P \cup E = R.$$

Thus, when P is a strict partial order and $R = P \cup E$ then the relation $I = R \cap R^*$ as defined from R is identical to E. However, if we begin with a preorder R and define $I = R \cap R^*$ and $P = R \cap R^{c*}$ then I is not necessarily the same as E defined from P. For example, suppose $X = \{x,y\}$ and not xRy & not yRx. Then x and y are incomparable and $(x, y) \notin I = R \cap R^*$. Moreover, not yPx and not xPy so that $(x, y) \in (P \cup P^*)^c$ from which it follows that $(x, y) \in E$.

The form of R that is the direct correspondent of a strict partial order has not been specified above. It is

011. R is negatively transitive and connected.

Suppose $R \in \{011\}$. Then, with $I = R \cap R^*$ and $P = R \cap R^{c*}$, P is a strict partial order and xIy if and only if not xPy & not yPx. Conversely, suppose P is a strict partial order. Then, with $I = (P \cup P^*)^c$ and $R = P \cup I$, $R \in \{011\}$ and xIy if and only if xRy & yRx. The proofs of these theorems are not especially difficult and should be an instructive exercise for the reader. R as in 011 has not been referred to as an 'order' since it is not necessarily transitive.

In addition, we note that a direct R correspondent of a suborder P arises when R is complete and the transitive closure R^{ct} of the complement of R is irreflexive.

The real-valued representation (with X countable) for a preorder corresponds to that for a strict partial order (02) with E therein replaced by $I = R \cap R^*$, and for a complete preorder we have xRy if and only if $f(x) \geqslant f(y)$.

CHAPTER 4

choice functions

If $\mathscr{X}$ is a nonempty set of nonempty sets then a function F on $\mathscr{X}$ such that $F(Y)$ is a *nonempty subset* of Y for each $Y \in \mathscr{X}$ is a *choice function.* $F(Y) \subseteq Y$ is the *choice set* from Y. When each $F(Y)$ contains exactly one element, then f on $\mathscr{X}$ defined by

$$F(Y) = \{f(Y)\} \quad \text{for} \quad \text{each } Y \in \mathscr{X}$$

is also called a choice function. Mirsky and Perfect (1966), who survey many interesting facts about choice functions, refer to f as a system of representatives. $f(Y)$ is the chosen element from Y or the *representative* of Y.

From the viewpoint of decision theory we might think of $f(Y)$ as the element an individual would choose from Y if he were required to make such a choice. $F(Y)$ can be viewed as the subset of 'best' elements from Y. Since it is entirely possible to construct sets that have no best element, some care may be required when considering the structure of $\mathscr{X}$. In what follows we shall assume that all sets in $\mathscr{X}$ are subsets of a set X and consider some theories based on properties of choice functions and structures for $\mathscr{X}$. In these theories choice functions rather than preferences are the primitives. Various notions of preference, generally referred to as 'revealed preference', may then be defined in terms of choice functions. However, there appear to be situations where no reasonable definition of preference in terms of choice functions applies and in such cases the generality of formulation offered by the choice function approach seems advantageous.

In the next section I shall consider the original version of revealed preference in the theory of consumer choice that is due to Samuelson (1938). This involves a choice function of the f variety with X the nonnegative orthant of n-dimensional Euclidean space. Thereafter we shall consider more general X sets and a sampling of theory for F functions. The final section discusses systems of choice functions and, among other things, presents the famous 'impossibility theorem' of Arrow (1963).

Revealed preference

To preface our discussion of revealed preference we shall consider some utility theory based on preference P as the primitive notion. Let $X = \{(x_1, \ldots, x_n) : x_i \geqslant 0 \text{ for } i = 1, \ldots, n\}$, the nonnegative orthant of Re^n. $x = (x_1, \ldots, x_n)$ is a *commodity bundle:* x_i is the quantity of the ith commodity. The following axioms for P on X are similar to those presented by Wold (1943).

W1. *P on X is a weak order.*
W2. *xPy whenever $x_i \geqslant y_i$ for all i and $x_i > y_i$ for some i.*
W3. *If xPy & yPz then there are numbers α and β such that*

$$0 < \alpha < 1 \;\;\&\;\; 0 < \beta < 1 \;\;\&\;\; \alpha x + (1-\alpha)zPy \;\;\&\;\; yP\beta x + (1-\beta)z.$$

It can be shown that {W1, W2, W3} is sufficient for the following proposition: there is a real-valued (utility) function u on X such that, for all $x, y \in X$,

$$xPy \text{ if and only if } u(x) > u(y).$$

Neither W2 nor W3 is completely necessary for this proposition.

Now suppose that the consumer has an income $m \geqslant 0$ to spend on these commodities and that $p_i > 0$ is the unit price of commodity i. $p = (p_1, \ldots, p_n)$ is the *price vector*. If the consumer buys $x \in X$, he will spend $p \cdot x = p_1x_1 + \ldots + p_nx_n$, which must not exceed m. Thus $[p, m]$ defined by $[p, m] = \{x : x \in X \;\&\; p \cdot x \leqslant m\}$, is his *opportunity set*. Assuming that W1 and W2 and W3 hold and that the consumer wishes to maximize his utility (or preference), it can be shown that

$$F(p, m) = \{x : x \in [p, m] \;\&\; u(x) \geqslant u(y) \text{ for all } y \in [p, m]\}$$

is not empty. According to W2, $p \cdot x = m$ for all $x \in F(p, m)$ so that the consumer spends all his income. Clearly, F is a choice function on $\mathscr{X} = \{[p, m] : p_i > 0 \text{ for all } i \;\&\; m \geqslant 0\}$. It is also referred to as a demand function.

If additional requirements are placed on P, we can insure that each $F(p, m)$ contains exactly one bundle. Letting $I = (P \cup P^*)^c$, the following axiom has this effect.

W4. *If* xIy & $x \neq y$ *then* $\frac{1}{2}x+\frac{1}{2}yPx$.

Suppose in fact that $x, y \in F(p, m)$ and $x \neq y$. Then $\frac{1}{2}x+\frac{1}{2}y \in [p, m]$ and since $\frac{1}{2}x+\frac{1}{2}yPx$ by W4, $u(\frac{1}{2}x+\frac{1}{2}y) > u(x)$, contradicting $x \in F(p, m)$. With $f(p, m)$ the unique bundle in $F(p, m)$, f on $\mathscr{X}$ is a choice function (in the second sense) that is also referred to as a demand function, with $f(p, m)$ the demand generated by p and m.

For $n = 2$ Figure 1 illustrates one case where each $[p, m]$ gives rise to a unique demand in X. Curve C is an indifference set that touches the budget line $p \cdot x = m$ at a unique point c, with $f(p, m) = c$.

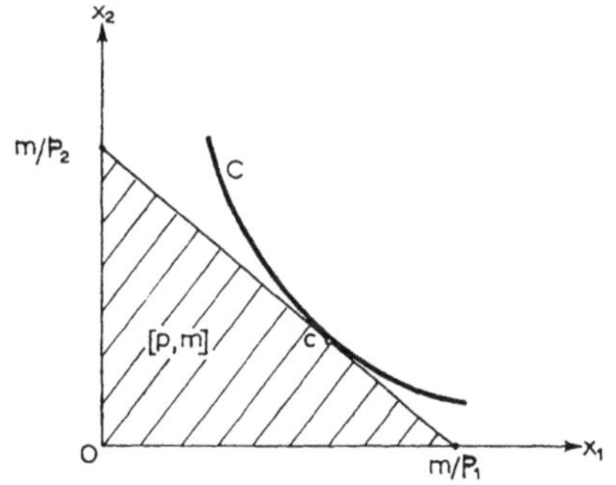

Figure 1. $u(x) = u(y)$ when both x and y are in C.

Samuelson's revealed preference approach (1938, 1948) came into being against a background such as this. He was interested in basing consumption theory on a demand function f on $\mathscr{X}$ rather than on preference as the primitive.

Suppose then that we begin with a choice or demand function f on $\mathscr{X} = \{[p,m] : p_i > 0 \text{ for all } i \;\&\; m \geq 0\}$. Since $f(p,m)$ is chosen rather than some other commodity bundle in the opportunity set $[p, m]$ we say that $f(p, m)$ is revealed to be preferred to each other bundle in $[p, m]$. P°, *defined* by

$$(x, y) \in P^\circ \quad \text{if and only if there is a} \quad [p, m] \in \mathscr{X} \quad \text{such that}$$
$$f(p, m) = x \quad \& \quad y \in [p, m] \quad \& \quad y \neq x,$$

is a *revealed preference relation.* Since Samuelson assumes that the total income m is spent on $f(p, m)$, or that $f(p, m) \cdot p = m$, the preceding definition is the same as

$$xP^{\circ}y \text{ if and only if there is a } [p, m] \in \mathscr{X} \text{ such that}$$

$$f(p, m) = x \quad \& \quad p \cdot y \leqslant p \cdot x \quad \& \quad y \neq x.$$

In this context Samuelson introduces what has since been referred to as his *weak axiom of revealed preference:*

WEAK AXIOM. *P° is asymmetric.*

Thus if $xP^{\circ}y$ and if $[p', m'] \in \mathscr{X}$ has $f(p', m') = y$ then $p' \cdot x > p' \cdot y$. Later Houthakker (1950) introduced his *strong axiom of revealed preference:*

STRONG AXIOM. *P° is a suborder,*

which means (Chapter 3) that the transitive closure $P^{\circ t}$ of P° is irreflexive. Under certain conditions the strong axiom can be shown to be implied by the weak axiom.

A good discussion of these two axioms along with additional theory is provided by Uzawa (1960) and Houthakker (1961). Both Houthakker (1950) and Uzawa (1960) establish sufficient conditions for the following proposition: there is a real-valued function u on X such that, for every $[p, m] \in \mathscr{X}$,

$$\{f(p, m)\} = \{x : x \in [p, m] \quad \& \quad u(x) \geqslant u(y) \quad \text{for all} \quad y \in [p, m]\}.$$

More recently Richter (1966) shows that the following axioms are sufficient for this proposition:

f1. *For each $x \in X$ there is a $[p, m] \in \mathscr{X}$ such that $f(p, m) = x$.*
f2. *$f(p, m) \cdot p = m$ for every $[p, m] \in \mathscr{X}$.*
f3. *P° is a suborder* (Strong Axiom).

The latest discussion along these lines is given by Hurwicz and Richter (1971).

Subset choice functions

In this section I shall present four developments based on subset choice functions F on a nonempty set $\mathscr{X}$ of nonempty subsets of an arbitrary nonempty set X. The first three of these involve conditions on F and the structure of $\mathscr{X}$ that imply the following proposition: there is a complete preorder R on X such that, for all $Y \in \mathscr{X}$,

$$F(Y) = \{x : x \in Y \quad \& \quad xRy \quad \text{for all} \quad y \in Y\}.$$

1. Arrow (1959) assumes that x contains every nonempty finite subset of X. One of his axioms is (for all $Y, Z \in \mathscr{X}$):

A1. *If* $Y \subseteq Z \;\&\; Y \cap F(Z) \neq \phi$ *then* $F(Y) = Y \cap F(Z)$.

This says that if Y contains an element in the choice set of a set that includes Y then the choice set for Y shall equal all such elements. Based on F, Arrow defines the following version of revealed preference:

$$(x, y) \in P' \text{ if and only if } x \in F(Y) \;\&\; y \in Y - F(Y) \text{ for some } Y \in \mathscr{X}.$$

He then introduces two axioms that are versions of the weak axiom (A2) and the strong axiom (A3).

A2. *If* $xP'y$ *then* $x \in Y \;\&\; y \in F(Y)$ *for no* $Y \in \mathscr{X}$.
A3. *If* $x, y \in F(Y)$ *for some* $Y \in \mathscr{X}$ *then* $(x, y) \notin P'^t$.

When $\mathscr{X}$ contains every finite subset of X, then A1, A2, and A3 are equivalent (each implies the others), and R on X defined by

$$(x, y) \in R \quad \text{if and only if} \quad x \in F(\{x, y\})$$

is a complete preorder that satisfies $F(Y) = \{x : x \in Y \;\&\; xRy \text{ for all } y \in Y\}$ provided that A1 holds.

2. Hansson (1968) considers conditions like A1, A2, and A3 without assuming that all finite subsets of X are in $\mathscr{X}$. He defines $(\mathscr{X}', F')$ as an *extension* of $(\mathscr{X}, F)$ if F' is a choice function on $\mathscr{X}'$, $\mathscr{X} \subseteq \mathscr{X}'$, and $F'(Y) = F(Y)$ for all $Y \in \mathscr{X}$. One of his theorems says that if A1 holds then the following three propositions are equivalent.

a. There is a complete preorder R on X such that $F(Y) = \{x : x \in Y \;\&\; xRy \text{ for all } y \in Y\}$ for all $Y \in \mathscr{X}$.
b. There is an extension $(\mathscr{X}', F')$ of $(\mathscr{X}, F)$ that satisfies A1 and is such that every nonempty finite union of sets in $\mathscr{X}'$ is in $\mathscr{X}'$.
c. There is an extension $(\mathscr{X}', F')$ of $(\mathscr{X}, F)$ that satisfies A1 and is such that every nonempty finite subset of X is in $\mathscr{X}'$.

3. Richter (1966) presents a necessary and sufficient condition for the existence of a complete preorder R on X ... that makes no demands on the structure of $\mathscr{X}$. Defining a version of revealed preference-or-indifference he takes

$$(x, y) \in R' \quad \text{if and only if} \quad x \in F(Y) \quad \& \quad y \in Y \quad \text{for some} \quad Y \in \mathscr{X}.$$

His condition, which uses the transitive closure of R', is

A4. *If* $x \in F(Y)$ & $y \in Y$ & $(y, x) \in R'^{t}$ *then* $y \in F(Y)$.

This says that if x is in the choice set of a set Y that contains y and if y is indirectly revealed preferred-or-indifferent to x (in the sense that $yR'^{t}x$) then y too is in the choice set. Richter's theorem says that A4 holds if and only if there is a complete preorder R on X such that $F(Y) = \{x : x \in Y \;\&\; xRy \text{ for all } y \in Y\}$ for all $Y \in \mathscr{X}$.

It follows from this theorem and that due to Arrow (1959) that A4 is equivalent to A1 when $\mathscr{X}$ includes all nonempty finite subsets of X. Regardless of the structure of $\mathscr{X}$, it is easily shown that, when A2 holds, $xP'y$ if and only if $xR'y$ and not $yR'x$, or $P' = R' \cap R'^{c*}$.

4. Arrow (1959) has two more conditions that are weaker than his other conditions. These are (for all $Y, Z \in \mathscr{X}$):

A5. *If* $Y \subseteq Z$ *then* $Y - F(Y) \subseteq Z - F(Z)$.

A6. *If* $Y \subseteq Z$ *then* $Y \cap F(Z) \subseteq F(Y)$.

These are easily shown to be equivalent regardless of the structure of $\mathscr{X}$. A5 says that if an element in Y is not in the choice set for Y and if Z includes Y then the element will not be in Z's choice set. The similarities and differences between A1 and A6 are obvious.

If $\mathscr{X}$ includes all nonempty finite subsets of X and P on X is defined by

$$(x, y) \in P \quad \text{if and only if} \quad x \neq y \quad \& \quad F(\{x, y\}) = \{x\}$$

then A5 implies that P is a suborder and that $F(Y) \subseteq \{x : x \in Y \;\&\; yPx \text{ for no } y \in Y\}$ for all $Y \in \mathscr{X}$.

Systems of choice functions

Some investigations in decision theory are based on interrelationships among choice functions in a set of choice functions each of which is defined on a set $\mathscr{X}$. Each choice function in the set is conditioned by a specific datum D

in a set $\mathcal{D}$ of possible alternative data points. Instead of writing the choice function for D as F_D, I shall use the equivalent device of writing $F(Y, D) = F_D(Y)$ and let F be a function on the product $\mathcal{X} \times \mathcal{D}$. The natural interpretation is that $F(Y, D)$ is the choice set from Y in the event that Y is the set of feasible alternatives and datum D obtains. We assume $\phi \subset F(Y, D) \subseteq Y$ for all $(Y, D) \in \mathcal{X} \times \mathcal{D}$.

Here are several contexts in which such a formulation might be adopted:

1. A decision maker will perform an experiment before making a decision. $\mathcal{D}$ is the set of possible outcomes of the experiment. If $D \in \mathcal{D}$ obtains and Y is the available set of basic alternatives he will then implement some $x \in F(Y, D)$.
2. An individual is to provide a preference order D_i on the alternatives in X under each of n different criteria $(i = 1, \ldots, n)$. $D = (D_1, \ldots, D_n)$ is an n-tuple of such preference orders. $\mathcal{D}$ is the set of all possible n-tuples.
3. Each of n individuals in a group is to provide a preference order on X. D_i is the order for individual i, $D = (D_1, \ldots, D_n)$, and $\mathcal{D}$ is the set of all allowable n-tuples of individual orders. $F(Y, D)$ is the group's choice set when Y is the available subset of X and D is the datum that obtains.

To illustrate the use of F on $\mathcal{X} \times \mathcal{D}$ I shall consider two contributions to the theory of social choice (context 3), although interpretations of the results noted below are not confined to this context.

Our first example is due to May (1952, 1953). Let $X = \{x, y\}$ with $x \neq y$. Throughout we shall let 1, 0, and -1 represent $\{x\}$, $\{x, y\}$, and $\{y\}$ respectively. $\mathcal{D} = \{1, 0, -1\}^n$ with $D = (D_1, \ldots, D_n)$ showing the choice of each of the n voters. $D_i = 0$ represents indifference or abstention. We shall let $\mathcal{X} = \{X\}$ and therefore omit X from $F(X, D)$.

F is a simple majority social choice function when, for each $D \in \mathcal{D}$, $F(D) = 1, 0,$ or -1 according to whether $\sum D_i > 0$, $\sum D_i = 0$, or $\sum D_i < 0$. May (1952) proves that the following conditions are necessary and sufficient for F to be a simple majority social choice function.

B1. (Anonymity). $F(D'_1, \ldots, D'_n) = F(D_1, \ldots, D_n)$ *if there is a permutation* σ *of* $\{1, \ldots, n\}$ *such that* $D_i = D'_{\sigma(i)}$ *for* $i = 1, \ldots, n$.
B2. (Duality). $F(-D_1, -D_2, \ldots, -D_n) = -F(D_1, D_2, \ldots, D_n)$.
B3. (Monotonicity). *If* $D'_i \geqslant D_i$ *for all* i & $D'_i > D_i$ *for some* i & $F(D) \geqslant 0$ *then* $F(D') = 1$.

The proof of May's theorem is not too difficult and is suggested to the interested reader as an exercise.

Our second example for F on $\mathcal{X} \times \mathcal{D}$ is due to Arrow (1963) and Blau (1957). For simplicity we shall assume that X is finite and let $\mathcal{X}$ be the set of all nonempty subsets of X. $\mathcal{D} = \{(D_1, \ldots, D_n) : D_i$ is a weak order on X for $i = 1, \ldots, n\}$. We consider the following conditions for all $Y, Z \in \mathcal{X}$ and $D, D' \in \mathcal{D}$. The first condition is A1 applied to each $D \in \mathcal{D}$.

C1. *If* $Y \subseteq Z$ & $Y \cap F(Z, D) \neq \phi$ *then* $F(Y, D) = Y \cap F(Z, D)$.

C2. (Independence of Irrelevant Alternatives). *If for all i and all* $x, y \in Y$, xD_iy *if and only if* xD'_iy, *then* $F(Y, D) = F(Y, D')$.

C3. (Unanimity). *If* xD_iy *for all i then* $F(\{x, y\}, D) = \{x\}$.

C4. (No Dictator). *There is no i such that, for all* $x, y \in X$ *and* $D \in \mathcal{D}$ *in which* xD_iy, $F(\{x, y\}, D) = \{x\}$.

Arrow proves that {C1, C2, C3, C4} is inconsistent when X contains at least three alternatives. His proof of this is divided into two parts. In the first part it is shown that if C1 & C2 & C3 hold and if there is an i and $x \neq y$ such that $F(\{x, y\}, D) = \{x\}$ when xD_iy and yD_jx for all $j \neq i$ then i is a dictator in the sense disallowed by C4. The second part of the proof then shows that C1 & C2 & C3 imply that there is such an i and $x \neq y$ .... Hence C1 & C2 & C3 imply the contradictory of C4.

Much of the furor and research that stemmed from the early published announcements of a theorem like this [Arrow (1950) and the 1951 edition of Arrow (1963)] is summarized by Arrow (1963, pp. 92–120). The significance of his 'impossibility theorem' is still an active issue today and continues to provide food for research thought.

CHAPTER 5

finite linear systems

In this chapter we shall be interested in systems of linear inequalities and equalities of the form

$$\varrho \cdot x > 0 \quad \text{for every} \quad x \in A$$

$$\varrho \cdot x = 0 \quad \text{for every} \quad x \in B$$

$$\varrho \cdot x \leqslant 0 \quad \text{for every} \quad x \in C$$

where each of A, B, and C is a finite subset of Re^N and A is not empty. Such a system has a *solution* if and only if there is a $\varrho \in Re^N$ that satisfies the specified linear inequalities and equalities.

The book edited by Kuhn and Tucker (1956), and especially the articles by Tucker and Goldman therein, provides basic theory for these and related systems. As we shall see, there are numerous applications of this theory in decision theory. I shall describe four such applications, identified here by decision context and the nature of B and C:

1. Expected utilities: $B = C = \phi$.
2. Semiorders: $B = C = \phi$.
3. Additive utilities: $C = \phi$.
4. Special majority social choice: $B = \phi$.

These applications will be considered after we present some basic solution theory for the systems described above.

Convex sets and solution vectors

A set $X \subseteq Re^N$ is *convex* if and only if $\lambda x + (1-\lambda)y \in X$ whenever $x, y \in X$ and $0 \leqslant \lambda \leqslant 1$.

Associated with A, B, and C as described above we shall consider the *convex closure* A' of A, the *vector space* B^v *generated by* B, and the *convex cone with origin* C'' *generated by* C. Each of these is a convex set, and with $A \neq \phi$ they are defined as follows, all sums being finite.

$$A' = \{\textstyle\sum_{x \in A}\lambda(x)\,x : \lambda(x) \geqslant 0 \quad \text{for all} \quad x \in A \;\&\; \sum_A \lambda(x) = 1\},$$
$$B^v = \{\textstyle\sum_{x \in B}\sigma(x)\,x : \sigma(x) \in Re \quad \text{for all} \quad x \in B\} \cup \mathbf{0},$$
$$C'' = \{\textstyle\sum_{x \in C}\mu(x)\,x : \mu(x) > 0 \quad \text{for all} \quad x \in C\} \cup \mathbf{0}.$$

Suppose $\varrho \cdot x > 0$ for all $x \in A$. Then $\varrho \cdot y > 0$ for all $y \in A'$ since $\varrho \cdot y = \varrho \cdot (\sum_A \lambda(x)x) = \sum_A \lambda(x)(\varrho \cdot x)$ and $\lambda(x) \geqslant 0$ for all $x \in A$, $\lambda(x) > 0$ for some $x \in A$, and $\varrho \cdot x > 0$. Similarly if $\varrho \cdot x = 0$ for all $x \in B$ then $\varrho \cdot y = 0$ for all $y \in B^v$; and if $\varrho \cdot x \leqslant 0$ for all $x \in C$ then $\varrho \cdot y \leqslant 0$ for all $y \in C''$.

Thus if our finite system has a solution when $C = \phi$ then $A' \cap B^v = \phi$. If our finite system has a solution when $B = \phi$ then $A' \cap C'' = \phi$. On the other hand, if the specified intersection is empty then it can be proved that the corresponding finite system has a ϱ solution. If there is no solution when $C = \phi$ it then follows that A' and B^v must contain a common point, so that $\sum_A \lambda(x)x = \sum_B \sigma(x)x$ or else $\sum_A \lambda(x)x = \mathbf{0}$. Similarly, if there is no solution when $B = \phi$ then $\sum_A \lambda(x)x = \sum_C \mu(x)x$ or else $\sum_A \lambda(x)x = \mathbf{0}$ for some λ's and μ's that satisfy the defining conditions. These facts are summarized in the following two lemmas where the r_k correspond directly to λ's, $-\mu$'s, and $-\sigma$'s that describe a common point when $A' \cap B^v \neq \phi$ or when $A' \cap C'' \neq \phi$. Since $\varrho \cdot x = 0$ if and only if $\varrho \cdot x \leqslant 0$ and $\varrho \cdot (-x) \leqslant 0$, Lemma 1 follows readily from Lemma 2.

LEMMA 1. *Suppose* $x^1, \ldots, x^M$ *are in* Re^N *and* $1 \leqslant K \leqslant M$. *Then either there is a* $\varrho \in Re^N$ *such that*

$$\varrho \cdot x^k > 0 \quad \textit{for} \quad k = 1, \ldots, K$$
$$\varrho \cdot x^k = 0 \quad \textit{for} \quad k = K+1, \ldots, M$$

or else there are nonnegative real numbers $r_1, \ldots, r_K$ *at least one of which is positive and real numbers* $r_{K+1}, \ldots, r_M$ *such that*

$$\textstyle\sum_{k=1}^{M} r_k x_i^k = 0 \quad \textit{for} \quad i = 1, \ldots, N.$$

LEMMA 2. *Suppose $x^1, \ldots, x^M$ are in Re^N and $1 \leqslant K \leqslant M$. Then either there is a $\varrho \in Re^N$ such that*

$$\varrho \cdot x^k > 0 \quad \textit{for} \quad k = 1, \ldots, K$$
$$\varrho \cdot x^k \leqslant 0 \quad \textit{for} \quad k = K+1, \ldots, M$$

or else there are nonnegative real numbers $r_1, \ldots, r_K$ at least one of which is positive and nonpositive real numbers $r_{K+1}, \ldots, r_M$ such that

$$\textstyle\sum_{k=1}^{M} r_k x_i^k = 0 \quad \textit{for} \quad i = 1, \ldots, N.$$

When $M = K$, so that $B = C = \phi$ and $B^{\text{v}} = C'' = \mathbf{0}$, Lemmas 1 and 2 are identical and we have the finite system $\{\varrho \cdot x > 0 : x \in A\}$. Figure 2-I illustrates this case in Re^2 with A containing four points. A' is the shaded polygon. Since $\mathbf{0} \notin A'$ we know that the system has a solution. To illustrate one such solution we pass a straight line H through $\mathbf{0}$ that does not intersect A'. The normal (perpendicular) to H at $\mathbf{0}$ is P. If ϱ is any point on P other than $\mathbf{0}$ then $H = \{x : \varrho \cdot x = 0\}$. If ϱ is on the same side of H as A'

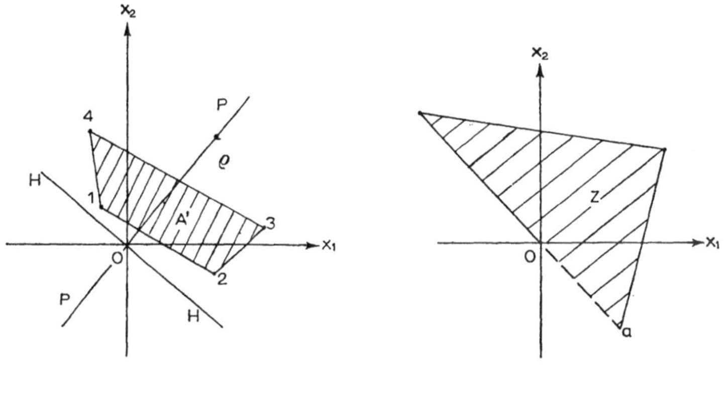

Figure 2. Convex sets

then $\varrho \cdot x > 0$ for all x on the same side of H as A'. Therefore $\varrho \cdot x > 0$ for all $x \in A'$ and hence for all $x \in A$.

Figure 2-II shows a convex set Z that includes all boundary points except for those points on the line segment from $\mathbf{0}$ to a inclusive. Although Z is the convex closure of many sets of points in Re^2, it is *not* the convex closure of any *finite* number of points in the plane. Although $\mathbf{0} \notin Z$ it is impossible to pass a straight line through $\mathbf{0}$ that does not intersect Z, and there is no $\varrho \in Re^2$ for which $\varrho \cdot x > 0$ for all $x \in Z$. Hence the theory given for $\{\varrho \cdot x > 0 : x \in A\}$ does not apply when A is infinite. Solution theory for this more general case is discussed by Klee (1955), Aumann (1962), and Kannai (1963).

In applying these lemmas to problems in decision theory we shall use a third lemma that will simplify things.

LEMMA 3. *Suppose that each x^k ($k = 1, \ldots, M$) is rational. Then if there is a ϱ solution in Lemma 1 or in Lemma 2 there is an integral ϱ solution: and if there is no ϱ solution in Lemma 1 or in Lemma 2 then there are integers $r_1, \ldots, r_M$ that satisfy $\sum_{k=1}^{M} r_k x^k = \mathbf{0}$ and have the other properties indicated in the applicable lemma.*

In the applications that follow, our main concerns will be to show how to formulate a situation for the use of the lemmas and how to make use of the lemmas once this formulation has been completed. We shall begin with a desired numerical system and obtain conditions that are sufficient (and also necessary under finiteness restrictions) for the numerical system.

Expected utilities

For our first case we shall let S be a finite set with $N > 0$ elements and let X be a nonempty finite set of probability distributions on S. Each $x \in X$ can be viewed as a real-valued function on S for which

$$x(s) \geqslant 0 \quad \text{for all} \quad s \in S, \quad \text{and} \quad \textstyle\sum_s x(s) = 1.$$

We shall take P (preference) on X and inquire about the existence of a real-valued function u on S such that, for all $x, y \in X$,

$$xPy \quad \text{implies} \quad \textstyle\sum_s x(s)u(s) > \sum_s y(s)u(s).$$

The sum $\sum x(s)\,u(s)$ is often referred to as the expected utility of the probability distribution x on S.

If $P = \phi$, any u on S satisfies the system. If $P \neq \phi$ we shall suppose that $P = \{(x_1, y_1), \ldots, (x_K, y_K)\}$. With $\varrho = (u(s_1), \ldots, u(s_N))$ with N components and $S = \{s_1, \ldots, s_N\}$ we shall let

$$a^k = (x_k(s_1) - y_k(s_1), \ldots, x_k(s_N) - y_k(s_N)).$$

Then $\sum_s x_k(s)\,u(s) > \sum_s y_k(s)\,u(s)$ is the same as $\varrho \cdot a^k > 0$.

If $\{\varrho \cdot a^k > 0 : k = 1, \ldots, K\}$ has no solution then, by Lemma 1, there are nonnegative numbers $r_1, \ldots, r_K$ with $r_k > 0$ for some k such that

$$\textstyle\sum_{k=1}^{K} r_k a^k = \mathbf{0}.$$

Without loss in generality we can suppose that $\sum r_k = 1$ so that we may interpret the r_k as probabilities. Since $\sum r_k a^k = \mathbf{0}$ is the same as $\sum r_k x_k(s) = \sum r_k y_k(s)$ for each $s \in S$, we consider the

AXIOM. *If $m > 1$, if $x_j, y_j \in X$ for $j = 1, \ldots, m$, if $r_j > 0$ for $j = 1, \ldots m$ and $\sum_j r_j = 1$, and if $\sum_j r_j x_j(s) = \sum_j r_j y_j(s)$ for each $s \in S$ then not $x_j P y_j$ for at least one $j \in \{1, \ldots, m\}$.*

Our preceding analysis shows that this Axiom implies the existence of a real-valued function u on S that satisfies our original expected-utility representation.

To interpret the Axiom we note that $(r_1x_1, \ldots, r_mx_m)$ can be viewed as a compound gamble or lottery. If $(r_1x_1, \ldots, r_mx_m)$ is 'chosen' then an x_j is chosen therefrom according to the 'probabilities' $r_1, \ldots, r_m$. The chosen x_j then gives rise to an $s \in S$ according to the probabilities $x_j(s_1), \ldots, x_j(s_N)$. The total probability that s will be chosen from $(r_1x_1, \ldots, r_mx_m)$ is $\sum_j r_j x_j(s)$.

Thus if $\sum_j r_j x_j(s) = \sum_j r_j y_j(s)$ for all $s \in S$ we would expect an individual to be indifferent between the compound lotteries $(r_1x_1, \ldots, r_mx_m)$ and $(r_1y_1, \ldots, r_my_m)$. However, if he prefers x_j to y_j for each $j \in \{1, \ldots, m\}$, we would expect him to prefer $(r_1x_1, \ldots, r_mx_m)$ to $(r_1y_1, \ldots, r_my_m)$. Therefore, our Axiom requires not $x_j P y_j$ for some j when $\sum r_j x_j(s) = \sum r_j y_j(s)$ for all s.

Semiorders

For our second example we shall examine the following proposition: there is a real-valued function u on X and a number $\delta > 0$ such that, for all $x, y \in X$,

$$xPy \quad \text{if and only if} \quad u(x) > u(y) + \delta.$$

It is easily seen that this implies that P is a semiorder so that P2, P10 and P11 hold:

P2. P is irreflexive.

P10. xPy & zPw imply xPw or zPy.

P11. xPy & yPz imply xPw or wPz.

We shall assume henceforth that X is a nonempty finite set and that P on X is a semiorder. Our problem is to show that these conditions imply the truth of the utility proposition. We shall follow Scott (1964).

If the representation holds and $(x, y) \notin P$ then $u(x) \leq u(y)+\delta$. If $=$ holds here we can replace it with $<$ by increasing δ slightly without altering the validity of $u(x) > u(y)+\delta$ when $(x, y) \in P$. Therefore we can write our system of inequalities as

$$u(x)-u(y)-\delta > 0 \quad \text{for all} \quad (x, y) \in P$$
$$u(y)-u(x)+\delta > 0 \quad \text{for all} \quad (x, y) \notin P.$$

Since $(x, x) \notin P$ by P2, the latter insures that $\delta > 0$ if the system has a solution.

To apply Lemma 1 with $B = \phi$ (or $K = M$) let N equal the size of X plus one and take $\varrho = (u(s), \ldots, u(t), \delta)$ where $s, \ldots, t$ exhaust X. For each $(x, y) \in P$ let a^k be such that $\varrho \cdot a^k = u(x)-u(y)-\delta$, so that $a_N^k = -1$. For each $(x, y) \notin P$ let a^k be such that $\varrho \cdot a^k = u(y)-u(x)+\delta$, so that $a_N^k = 1$. Letting K_1 be the number of $(x, y) \in P$ and K_2 be the number of $(x, y) \notin P$, our system can be written as

$$\varrho \cdot a^k > 0 \quad \text{for} \quad k = 1, \ldots, K_1 \qquad (\text{each } a_N^k = -1)$$
$$\varrho \cdot a^k > 0 \quad \text{for} \quad k = K_1+1, \ldots, K_1+K_2 \qquad (\text{each } a_N^k = 1)$$

For Lemma 1, $K = M = K_1+K_2$. If this system has no ϱ solution then we know from Lemmas 1 and 3 that there are nonnegative integers $r_1, \ldots, r_K$ at least one of which is positive such that

$$\textstyle\sum_{k=1}^{K} r_k a_i^k = 0 \quad \text{for} \quad i = 1, 2, \ldots, N.$$

Supposing that there is no ϱ solution, the properties for a_N^k along with $r_1 a_N^1 + \ldots + r_K a_N^K = 0$ imply that $\sum_{k=1}^{K_1} r_k = \sum_{k=K_1+1}^{K} r_k = T > 0$. It therefore follows from the definition of the a^k and from $\sum r_k a^k_i = 0$ for $i = 1, \ldots, N-1$ that there are sequences

$$x_1, \ldots, x_T, z_1, \ldots, z_T \quad \text{and} \quad y_1, \ldots, y_T, w_1, \ldots, w_T$$

such that $(x_k, y_k) \in P$ and $(w_k, z_k) \notin P$ for $k = 1, \ldots, T$ (with the possibility that $z_k = w_k$ for some k) and such that the sequence $y_1, \ldots, y_T, w_1, \ldots, w_T$ is a rearrangement (permutation) of the sequence $x_1, \ldots, x_T, z_1, \ldots, z_T$.

The final task in the proof is to show that the existence of such sequences contradicts the assumption that P is a semiorder. The reader is referred to Scott (1964, pp. 240–241) for this demonstration. From the contradiction we then know that $\sum r_k a^k = \mathbf{0}$ must be false and therefore know that the system has a ϱ solution.

Additive utilities

For our third example we shall assume that $X \subseteq X_1 \times X_2 \times \ldots \times X_n$ and that, for all i, each $x_i \in X_i$ is the ith component of some n-tuple $x \in X$. We assume further that X is a nonempty finite set. With $x = (x_1, \ldots, x_n)$ and $y = (y_1, \ldots, y_n)$, we are interested in the following additive utility proposition: for each $i \in \{1, \ldots, n\}$ there is a real-valued function u_i on X such that, for all $x, y \in X$,

$$xPy \quad \text{if and only if} \quad u_1(x_1) + \ldots + u_n(x_n) > u_1(y_1) + \ldots + u_n(y_n).$$

If this holds then P on X must be a weak order and, with $I = (P \cup P^*)^c$,

$$xIy \quad \text{if and only if} \quad u_1(x_1) + \ldots + u_n(x_n) = u_1(y_1) + \ldots + u_n(y_n).$$

To apply Lemma 1 we shall let $N = N_1 + \ldots + N_n$, where N_i is the number of elements in X_i, and take

$$\varrho = (u_1(x_{11}), \ldots, u_1(x_{1N_1}), u_2(x_{21}), \ldots, u_n(x_{nN_n}))$$

where, for each i, $\{x_{i1}, \ldots, x_{iN_i}\} = X_i$. Each $(x, y) \in P$ statement then translates into $\varrho \cdot a^k > 0$ where $\varrho \cdot a^k = u_1(x_1) + \ldots + u_n(x_n) - u_1(y_1) - \ldots - u_n(y_n)$ with each $a^k_i \in \{-1, 0, 1\}$ and $\sum_i a^k_i = 0$. Each $(x, y) \in I$ statement translates into $\varrho \cdot a^k = 0$ where $\varrho \cdot a^k = u_1(x_1) + \ldots + u_n(x_n) - u_1(y_1) - \ldots - u_n(y_n)$. Let K be the number of $(x, y) \in P$, and let $M-K$ be half the number of

$(x, y) \in I$ with $x \neq y$. In the following system we include only one of the two $\varrho \cdot a^k = 0$ statements from $(x, y) \in I$ and $(y, x) \in I$ for each $x \neq y$ since the other $\varrho \cdot a^k = 0$ statement is implied by the first. Thus our system is

$$\varrho \cdot a^k > 0 \quad \text{for} \quad k = 1, \ldots, K$$
$$\varrho \cdot a^k = 0 \quad \text{for} \quad k = K+1, \ldots, M$$

and our utility proposition holds if and only if this system has a solution.

Suppose that there is no ϱ solution. Then, by Lemmas 1 and 3 there are nonnegative integers $r_1, \ldots, r_K$ at least one of which is positive, and integers $r_{K+1}, \ldots, r_M$ such that

$$\textstyle\sum_{k=1}^{M} r_k a_i^k = 0 \quad \text{for} \quad i = 1, \ldots, N.$$

Suppose that $r_k < 0$ for some $k > K$. Then we can replace this r_k by its negative and simultaneously replace a^k by $-a^k$ in our system since $\varrho \cdot a^k = 0$ if and only if $\varrho \cdot (-a^k) = 0$ and $r_k a_i^k = (-r_k)(-a_i^k)$. Hence we may without loss in generality suppose that $r_k \geq 0$ for all k. After this replacement is made our system is

$$\varrho \cdot a^k > 0 \quad \text{for} \quad x^k P y^k \quad \text{with} \quad k = 1, \ldots, K$$
$$\varrho \cdot a^k = 0 \quad \text{for} \quad x^k I y^k \quad \text{with} \quad k = K+1, \ldots, M$$

with $\sum r_k a^k = \mathbf{0}$, $r_k \geq 0$ for all k and $r_k > 0$ for some $k \leq K$. Now with $m = \sum_{k=1}^{M} r_k$, $\sum r_k a^k = \mathbf{0}$ implies that there are sequences $x^1, \ldots, x^m$ and $y^1, \ldots, y^m$ of n-tuples in X such that, for each i,

$$y_i^1, \ldots, y_i^m \quad \text{is a rearrangement of} \quad x_i^1, \ldots, x_i^m.$$

That is, for any specific i there is a permutation σ of $\{1, \ldots, m\}$ such that $y_i^{\sigma(j)} = x_i^j$ for $j = 1, \ldots, m$. In addition, $x^j P y^j$ for $j = 1, \ldots, m'$ since $r_k > 0$ for some $k < K$, and $x^j I y^j$ for $j = m'+1, \ldots, m$. If $m = 1$ then $y^1 = x^1$ and we have $x^1 P x^1$ which is contradictory to irreflexivity.

This process reveals a sufficient condition for the existence of a solution for our finite system. Consider the

AXIOM. *If $m > 1$, if $x^j, y^j \in X$ for $j = 1, \ldots, m$, if for each i from 1 to n the sequence $y_i^1, \ldots, y_i^m$ is a rearrangement of $x_i^1, \ldots, x_i^m$ and if $x^j P y^j$ or $x^j I y^j$ for $j = 2, \ldots, m$ then $(x^1, y^1) \notin P$.*

As we have just seen, if there is no solution for the system and $\sum r_k > 1$ then the Axiom is contradicted. Moreover, with $m = 2$ and $x^1 = x^2 = y^1 = y^2 = x$, the Axiom shows that P is irreflexive, so that $\sum r_k = 1$ cannot hold when the Axiom holds. Therefore the Axiom implies that the finite system has a solution. Conversely, it is easily shown that the Axiom holds when the system has a solution. Hence, under the finiteness condition, additive utilities exist as described if and only if the Axiom holds.

Axioms that are similar to the one described here are discussed by Tversky (1964), Scott (1964), and Adams (1965).

Special majority

Our final example illustrates Lemma 2. We shall use the $1, 0, -1$ notation of the simple majority example in the final section of Chapter 4. Each $D \in \{1,0,-1\}^n$ represents an n-tuple of preferences for n individuals and, for each such D, $F(D)$ equals 1,0, and -1 according to whether the social choice is to be x, a tie between x and y, or y.

For each $D \in \{1, 0, -1\}^n$ let D^+ be the number of $D_i = 1$ (number of votes for x) and let D^- be the number of $D_i = -1$ (number of votes for y). F is a *special majority* social choice function if and only if there is a number $\beta \geq 0$ such that, for all $D \in \{1, 0, -1\}^n$,

$$F(D) = 1 \quad \text{if and only if} \quad D^+ > \beta D^-$$
$$F(D) = -1 \quad \text{if and only if} \quad D^+ \leq \beta D^-.$$

Under this F a tie, or $F(D) = 0$, never arises, and $F(1, \ldots, 1) = 1$ and $F(-1, \ldots, -1) = -1$. If y is viewed as the status quo and x is the challenger, then a two-thirds majority with abstentions ignored arises when $\beta = 2$ or when β is slightly less than 2.

The following three conditions are necessary and sufficient for the existence of a special majority social choice function.

C1. *$F(D) = 0$ for no $D \in \{1, 0, -1\}^n$.*

C2. *$F(0) = F(-1, \ldots, -1) = -1$ and $F(1, \ldots, 1) = 1$.*

C3. *If each of $D^1, \ldots, D^S, E^1, \ldots, E^T$ is in $\{1, 0, -1\}^n$ with $S \geq 1, T \geq 1$, and*

$$\textstyle\sum_{k=1}^S D^{k+} = \sum_{k=1}^T E^{k+} \quad \& \quad \sum_{k=1}^S D^{k-} = \sum_{k=1}^T E^{k-}$$

then $F(E^k) \geq 0$ for some E^k if $F(D^k) \geq 0$ for every D^k.

According to C1 & C2, $\mathcal{D} = \{1, 0, -1\}^n$ can be partitioned into $\mathcal{D}^+ = \{D : F(D) = 1\}$ and $\mathcal{D}^- = \{D : F(D) = -1\}$. Taking $N = 2$ in Lemma 2 we note that if the system

$$\varrho_1 D^+ + \varrho_2 D^- > 0 \quad \text{for all} \quad D \in \mathcal{D}^+$$

$$\varrho_1 D^+ + \varrho_2 D^- \leqslant 0 \quad \text{for all} \quad D \in \mathcal{D}^-$$

has a ϱ solution in which $-\varrho_2/\varrho_1 \geqslant 0$ then F is a special majority social choice function with $\beta = -\varrho_2/\varrho_1$. If this system has no ϱ solution then, by Lemmas 2 and 3, there are nonnegative integers $\{r(D) : D \in \mathcal{D}^+\}$ at least one of which is positive and nonnegative integers $\{r(D) : D \in \mathcal{D}^-\}$ such that

$$\textstyle\sum_{\mathcal{D}^+} r(D)\, D^+ - \sum_{\mathcal{D}^-} r(D)\, D^+ = 0$$

$$\textstyle\sum_{\mathcal{D}^+} r(D)\, D^- - \sum_{\mathcal{D}^-} r(D)\, D^- = 0.$$

If $r(D) > 0$ for some $D \in \mathcal{D}^-$ we then have a contradiction of C3. If $r(D) = 0$ for every $D \in \mathcal{D}^-$ then there is a $D \in \mathcal{D}^+$ with $D^+ = D^- = 0$ and hence $0 \in \mathcal{D}^+$, which contradicts C2. Hence there is a ϱ solution if C1, C2, and C3 hold. Moreover, C2 shows that $\varrho_1 > 0$ and that $\varrho_2 \leqslant 0$ so that $-\varrho_2/\varrho_1 \geqslant 0$

CHAPTER 6

Zorn's lemma

In this chapter we shall illustrate the use of Zorn's lemma. Properly considered, Zorn's lemma is an axiom of set theory. Some mathematicians have opposed the use of this axiom, which is independent of the other axioms of conventional set theory, but today most mathematicians do not hesitate to use it. Informally, Zorn's lemma says that if $\prec$ is a strict partial order on a set X and if each linearly ordered subset of X has an upper bound in X then there is a $\prec$-maximal element in X. A precise statement follows.

ZORN'S LEMMA. *Suppose that $\prec$ on X is a strict partial order and that, for each nonempty $Y \subseteq X$ for which the restriction of $\prec$ on Y is a linear order, there is an $x(Y) \in X$ such that $y \prec x(Y)$ or $y = x(Y)$ for all $y \in Y$. Then there is an $x^* \in X$ such that $x^* \prec x$ for no $x \in X$.*

Kelley (1955) shows that a number of other propositions are equivalent to Zorn's lemma. One of these is the

AXIOM OF CHOICE. *If $\mathscr{X}$ is a nonempty set of nonempty sets then there is a function f on $\mathscr{X}$ for which $f(Y) \in Y$ for each $Y \in \mathscr{X}$.*

In terms of Chapter 4, this says that there is a system of representatives for every such $\mathscr{X}$. Another equivalent proposition is the

WELL-ORDERING PRINCIPLE. *If X is a nonempty set then there is a linear order $\prec$ on X such that, for each nonempty $Y \subseteq X$, there is an $x(Y) \in Y$ for which $x(Y) \prec y$ or $x(Y) = y$ for every $y \in Y$.*

One example of the use of the Well-Ordering Principle is presented by Chipman (1960). We shall note this in the next chapter.

Szpilrajn's theorem

One of the better-known uses of Zorn's lemma arises in the proof of Szpilrajn's extension theorem (1930). In part, his theorem says that if P on X is a strict partial order then there is a linear order P' on X such that $P \subseteq P'$.

Suppose that P on $X \neq \phi$ is a strict partial order, and let $\mathcal{P}$ denote the set of all strict partial orders Q on X such that $P \subseteq Q$. Clearly, $\subset$ on $\mathcal{P}$ is a strict partial order. We apply Zorn's lemma to $\subset$ on $\mathcal{P}$. Let $\mathcal{R}$ be any subset of $\mathcal{P}$ that is linearly ordered by $\subset$. We wish to show that there is a $T \in \mathcal{P}$ such that $R \subset T$ or $R = T$ for all $R \in \mathcal{R}$. For this let

$$T = \cup_{\mathcal{R}} R,$$

so that $(x, y) \in T$ if and only if $(x, y) \in R$ for some $R \in \mathcal{R}$. By definition, $R \subseteq T$ for all $R \in \mathcal{R}$. It remains to show that $T \in \mathcal{P}$, that is, that T is a strict partial order. Since $(x, x) \notin R$ for all $R \in \mathcal{R}$, $(x, x) \notin T$ so that T is irreflexive. If $(x, y) \in T$ and $(y, z) \in T$ then $(x, y) \in R_1$ and $(y, z) \in R_2$ for some R_1, $R_2 \in \mathcal{R}$. Since $\mathcal{R}$ is linearly ordered by $\subset$, either $R_1 \subseteq R_2$ or $R_2 \subseteq R_1$. This and transitivity for R_1 and R_2 imply that $(x, z) \in R_1 \cup R_2$. Hence $(x, z) \in T$, and therefore $T \in \mathcal{P}$. Zorn's lemma then implies that there is a $P' \in \mathcal{P}$ such that $P' \subset R$ for no $R \in \mathcal{P}$. From the maximality of P' it is not hard to show that P' is weakly connected and is therefore a linear order.

Additive utility

Whereas the foregoing illustration uses Zorn's lemma in extending a given structure, our next illustration is concerned with obtaining a restriction of a given system. This application arose from an investigation of additive utilities for ordered pairs, details of which can be found in Fishburn (1970, Section 11.2). I shall only mention the part of this that involves the direct application of Zorn's lemma.

We assume that $X \subseteq X_1 \times X_2$ with each $x_1 \in X_1$ the first component of some $x \in X$ and each $x_2 \in X_2$ the second component of some $x \in X$. X can be finite or infinite. An equivalence E on X is defined as follows: xEy if and only if there is a finite sequence $x, x^1, \ldots, x^N, y$ of elements in X such that each two adjacent elements have a common component. E

partitions X into equivalence classes: if (x_1, x_2) and (y_1, y_2) in X are in different classes in X/E then $x_1 \neq y_1$ and $x_2 \neq y_2$. Thus the different classes in X/E are 'totally separate' from one another.

Let D be an arbitrary class in X/E. An *alternating cycle* in D is a finite subset of D that contains an even number n of elements that can be arranged in a sequence $x^1, x^2, \ldots, x^n$ so that:

a. each two adjacent elements have one common component,

b. no three consecutive elements have a component common to all three,

c. $x_1^n = x_1^1$ if $x_2^1 = x_2^2$, and $x_2^n = x_2^1$ if $x_1^1 = x_1^2$.

We shall use Zorn's lemma to show that $D \in X/E$ contains a subset C that has no cycles but retains other properties of D. In particular we wish to show that there is a $C \subseteq D$ such that:

A1. C includes no alternating cycle,

A2. xEy for each $x, y \in C$ is established by a sequence all of whose elements are in C,

A3. $(x_1, x_2) \in D$ implies that $(y_1, x_2) \in C$ for some $y_1 \in X_1$ and that $(x_1, y_2) \in C$ for some $y_2 \in X_2$. (This ensures that each element of X_1 and X_2 that appears in D appears in C also.)

Thus let $\mathcal{C} = \{C : C \subseteq D \text{ and } C \text{ satisfies A1 \& A2}\}$. We use Zorn's lemma to show that some $C \in \mathcal{C}$ satisfies A3. Let $\mathcal{C}'$ be a subset of $\mathcal{C}$ that is linearly ordered by $\subset$ and let

$$C' = \bigcup_{\mathcal{C}'} C, \quad \text{so that} \quad C \subseteq C' \quad \text{for all} \quad C \in \mathcal{C}'.$$

For Zorn's lemma we need to show that C' satisfies A1 and A2.

A1. If $\{x^1, \ldots, x^n\} \subseteq C'$ is a cycle then, since $\mathcal{C}'$ is linearly ordered by $\subset$, $\{x^1, \ldots, x^n\} \subseteq C$ for some $C \in \mathcal{C}'$. Since this contradicts A1 for C, there are no cycles in C'.

A2. If $x, y \in C'$ then $x, y \in C$ for some $C \in \mathcal{C}'$ so that xEy is established by elements in $C \subseteq C'$. Hence C' satisfies A2.

Thus, by Zorn's lemma, there is a $C^* \in \mathcal{C}$ such that $C^* \subset C$ for no $C \in \mathcal{C}$. Suppose that $(x_1, x_2) \in D - C^*$. Then, since C^* is maximal, $C^* \cup \{(x_1, x_2)\}$

must include a cycle that contains (x_1, x_2). It then follows from the definition of an alternating cycle that there are $(x_1, y_2) \in C^*$ and $(y_1, x_2) \in C^*$. This shows that C^* satisfies A3 as well as A1 and A2.

An unusual probability measure

Let $\mathcal{A}$ be the set of all subsets of a nonempty set X. A *probability measure* on $\mathcal{A}$ is a real-valued function p on $\mathcal{A}$ that satisfies

B1. $p(A) \geq 0$ for all $A \in \mathcal{A}$,

B2. $p(X) = 1$,

B3. $p(A \cup B) = p(A)+p(B)$ whenever $A, B \in \mathcal{A}$ and $A \cap B = \phi$.

We shall outline the steps that show that if X is an infinite set then there is a probability measure p on $\mathcal{A}$ that satisfies

B4. $p(A) \in \{0, 1\}$ for every $A \in \mathcal{A}$,

B5. $p(\{x\}) = 0$ for every $x \in X$.

With X infinite let H be the set of all sets $\mathcal{B}$ of subsets of X that have the following four properties:

1. $\phi \in \mathcal{B}$ and $\{x\} \in \mathcal{B}$ for all $x \in X$,
2. if $A \in \mathcal{B}$ then $X - A \notin \mathcal{B}$,
3. if $A, B \in \mathcal{B}$ then $A \cup B \in \mathcal{B}$,
4. if $B \subseteq A$ and $A \in \mathcal{B}$ then $B \in \mathcal{B}$.

Since the set of all finite subsets of X is such a $\mathcal{B}$, H is not empty. Zorn's lemma is then used to show that H contains a $\mathcal{B}^*$ such that $\mathcal{B}^* \subset \mathcal{B}$ for no $\mathcal{B} \in H$. This use of the lemma is easy and is suggested as an exercise.

With $\mathcal{B}^* \in H$ maximal in H under inclusion $\subset$, let $\mathcal{C}^*$ be the complement of $\mathcal{B}^*$ in $\mathcal{A}: \mathcal{C}^* = \{A : A \subseteq X \;\&\; A \notin \mathcal{B}^*\}$. From the maximality of $\mathcal{B}^*$ it can be shown that if $A, B \in \mathcal{C}^*$ then $A \cap B \neq \phi$. Using this fact along with properties 1 through 4 for $\mathcal{B}^*$ it can then be shown that p on $\mathcal{A}$ defined by

$$p(A) = 0 \quad \text{if} \quad A \in \mathcal{B}^*$$
$$p(A) = 1 \quad \text{if} \quad A \in \mathcal{C}^*$$

satisfies B1 through B5.

To indicate an application of such a p let V be an infinite set with p a probability measure on the set of all subsets of V such that $p(U) \in \{0, 1\}$ for all $U \subseteq V$ and $p(\{v\}) = 0$ for all $v \in V$. Each $v \in V$ may be thought of as an individual (voter) or as a criterion (performance characteristic, attribute). Suppose that $P(v)$ is a weak order on a set X of alternatives for each $v \in V$ and, for all $x, y \in X$, take

$$xPy \quad \text{if and only if} \quad p(\{v : xP(v)y\}) = 1.$$

It is not hard to prove that this P is a weak order on X. Moreover since $p(\{v : xP(v)y\}) = 1$ if $xP(v)y$ for all $v \in V$, P satisfies a unanimity property; and since $p(\{v : xP(v)y\}) = 0$ if $xP(v)y$ for only one $v \in V$, no one v dictates the collective preference P.

In the language of social choice theory each $U \subseteq V$ for which $p(U) = 1$ is a decisive set of voters: xPy whenever $xP(v)y$ for every $v \in U$. Under the given p there is no minimal decisive set since if $p(U) = 1$ and $v \in U$ then $p(U - \{v\}) = 1$.

We shall return to probability measures in Chapter 9, at which time we shall use the common notational abbreviation $p(x)$ for $p(\{x\})$.

CHAPTER 7

real-valued order-preserving functions

We have examined some of the theory of real-valued order-preserving functions for finite qualitative preference systems. We now begin an examination of such functions for sets that may be infinite. Succeeding chapters continue the theme of the present chapter for specially structured sets and special kinds of real-valued functions defined on these sets.

An example

If P on X is a weak order and $X/(P \cup P^*)^c$ is countable then there is a real-valued function u on X such that, for all $x, y \in X$, xPy if and only if $u(x) > u(y)$. [See, for example, Birkhoff (1948, p. 31), Suppes and Zinnes (1963, pp. 26–28), or Fishburn (1970, Theorem 2.2).] However, if X/I is uncountable, such a u may not exist.

To illustrate this let $X = Re^2$ with P a lexicographic linear order on X defined by $(x_1, x_2)P(y_1, y_2)$ if and only if $x_1 > y_1$ or $(x_1 = y_1$ & $x_2 > y_2)$. Suppose that there is a real-valued function u on X such that xPy if and only if $u(x) > u(y)$. Let $J(x_1) = [u(x_1, 0), u(x_1, 1)]$ for each $x_1 \in Re$. Each $J(x_1)$ is a nondegenerate closed interval and $J(x_1) \cap J(y_1) = \phi$ if $x_1 \neq y_1$. Since each $J(x_1)$ contains a rational number and there is an uncountable number of $x_1 \in Re$, we conclude that the set of rationals is uncountable, which is false. Hence there is no such u.

Despite this negative result any weak order can be represented by real numbers or by general vectors of real numbers. $(T, \prec)$ is a *well-ordered set* when $\prec$ is a linear order on T and each nonempty subset of T has a first element: if $\phi \subset S \subseteq T$ then there is an $s \in S$ such that $s \prec t$ for all $t \in S - \{s\}$.

Suppose that u is a real-valued function on $X \times T$ and $(T, \prec)$ is a (nonempty) well-ordered set. The set $\bar{u}(x) = \{(t, u(x, t)) : t \in T\}$ can be viewed as a vector of real numbers whose components, the $u(x, t)$, are ordered according to $\prec$. Thus, the first component of $\bar{u}(x)$ is $u(x, s)$ where $s \prec t$ for all $t \in T - \{s\}$. The second component of $\bar{u}(x)$ is $u(x, s')$ where $s' \prec t$ for all $t \in T - \{s, s'\}$; and so forth.

Relative to $(T, \prec)$ we define a lexicographic weak order $\overset{L}{>}$ on $\{\bar{u}(x) : x \in X\}$ as follows: $\bar{u}(x) \overset{L}{>} \bar{u}(y)$ if and only if $u(x, t) > u(y, t)$ for some $t \in T$ and, for each $t \in T$ for which $u(x, t) < u(y, t)$ there is an $s \prec t$ such that $u(x, s) > u(y, s)$. Using the Well-Ordering Principle to ensure the existence of a well-ordering for the set X/I, Chipman (1960) shows that if P on X is a weak order then there is a well-ordered set $(T, \prec)$ and a real-valued function u on $X \times T$ such that, for all $x, y \in X$,

$$xPy \quad \text{if and only if} \quad \bar{u}(x) \overset{L}{>} \bar{u}(y).$$

Two-dimensional vectors suffice for the preceding Re^2 example, with $\bar{u}(x) = (x_1, x_2)$ for all $x \in Re^2$. In the next section we present a condition that is necessary and sufficient for one-dimensional utility, or for T to be a unit set.

Order denseness

If P is a linear order on X then $Y \subseteq X$ is *P-order dense* (or simply order dense) in X if and only if, whenever xPz and $x, z \in X - Y$, there is a $y \in Y$ such that $xPyPz$. Thus the countable set of rational numbers is $>$-order dense in Re.

More generally, if P on X is a strict partial order we shall say that $Y \subseteq X$ is P-order dense in X if and only if, whenever xPz and $x, z \in X - Y$, it is false that yPx or zPy for each $y \in Y$. When P is a weak order, this says that if xPz and $x, z \in X - Y$ then $x(P \cup I)y(P \cup I)z$ for some $y \in Y$.

In the remainder of this chapter we shall comment on the nature and applications of the following

THEOREM. *There is a real-valued function u on X such that, for all $x, y \in X$,*

$$xPy \quad \textit{if and only if} \quad u(x) > u(y),$$

if and only if P on X is a weak order and there is a countable subset of X that is P-order dense in X.

I shall refer to this as Cantor's Theorem since it was first proved in a slightly different form by Cantor (1895). More recent proofs are given by Birkhoff (1948), Debreu (1954, 1964), Luce and Suppes (1965) and Fishburn (1970). The original proof is available in English in Cantor (1915), and Newman and Read (1961) contains useful discussion.

To indicate the sufficiency proof of Cantor's Theorem suppose P on X is a weak order and Y is a countable subset of X that is P-order dense in X. Define P' on X/I by $aP'b$ if and only if xPy for some (and hence for all) $x \in a$ & $y \in b$. P' on X/I is a linear order and $Y' = \{a : a \in X/I \;\&\; y \in a$ for some $y \in Y\}$ is P'-order dense in X/I. Let

$$Q = \{(a, b) : aP'b \text{ and there is no } c \in X/I \text{ such that } aP'cP'b\}.$$

If $(a, b) \in Q$ then either $a \in Y'$ or $b \in Y'$. Hence Q is countable. Therefore Y' along with all first and second components in the pairs in Q along with the best and worst element(s) under P' (if any) forms a countable set. Call this set Y^*. From the countability of Y^* we know that there is a real-valued function v on Y^* such that, for all $a, b \in Y^*$, $aP'b$ if and only if $v(a) > v(b)$. Defining $v(c)$ for each $c \in X/I - Y^*$ by

$$v(c) = \sup\{v(a) : a \in Y^* \;\&\; cP'a\}$$

it is not hard to see that, for all $a, b \in X/I$, $aP'b$ if and only if $v(a) > v(b)$. On defining $u(x) = v(a)$ when $x \in a$, it follows that, for all $x, y \in X$, xPy if and only if $u(x) > u(y)$.

In this proof we first defined our real-valued function on an appropriate countable dense subset and then extended it to the entire set (X/I) by squeezing the utilities for the other elements in between the utilities already defined. Instead of the supremum extension we could have used the infimum extension $v(c) = \inf\{v(a) : a \in Y^* \;\&\; aP'c\}$ and obtained an acceptable although possibly different v. Any convex combination of the sup and inf extensions also yields an order-preserving v.

Applications of Cantor's theorem

Cantor's Theorem is often useful in proving related theorems. For one case of this suppose that axioms W1, W2, and W3 of Chapter 4 hold for a set X where $X = \prod_{i=1}^{n} X_i$ and each X_i is a nonempty interval (open or closed or half-open) in Re. It can then be shown that, with $Y_i = \{r : r \in X_i$ and r_i is rational$\} \cup \{a : a \in X_i \,\&\, a \preccurlyeq b$ for all $b \in X_i\} \cup \{a : a \in X_i$ & $a \succcurlyeq b$ for all $b \in X_i\}$, the subset $Y = \Pi Y_i$ of X is P-order dense in X. Since each Y_i is countable, Y is countable and therefore Cantor's Theorem applies. For an explicit proof of order denseness in this case see Fishburn (1970, Section 3.3).

For a second application suppose that P on a nonempty set X is a semiorder. With $I = (P \cup P^*)^c$ we noted in Chapter 4 that $PI \cup IP$ is a weak order. We noted also that if X is finite then a closed unit real interval $K(x)$ can be assigned to each $x \in X$ so that, for all $x, y \in X$, xPy if and only if $\inf K(x) > \sup K(y)$. But there may be no such representation when X is infinite.

We now ask: What conditions for a semiorder P on an arbitrary nonempty set X are necessary and sufficient for the existence of a function J on X into the set of closed real intervals such that, for all $x, y \in X$, xPy if and only if $\inf J(x) > \sup J(y)$? Using Cantor's Theorem among other things it can be shown that the following two conditions are precisely those requested:

C1. *X includes a countable subset that is $PI \cup IP$-order dense in X.*

C2. *With $Q' = \{(x, y) : xPy \,\&\, (xPr \,\&\, rIs \,\&\, sPy)$ for no $r, s \in X\}$ and E' the equivalence defined on Q' by $(x, y)E'(z, w)$ if and only if $(x, y), (z, w) \in Q' \,\&\, xPw \,\&\, zPy$, Q'/E' is countable.*

A composite application

I conclude this chapter by outlining a proof, due in large part to Richter (1966), that involves not only Cantor's Theorem but also Szpilrajn's extension theorem and a number of other concepts introduced earlier in this book. For this reason the following may tell the reader how well he has grasped the material presented to this point.

We shall let X be the nonnegative orthant of Re^n and define xDy if and only if $x_i > y_i$ for $i = 1, \ldots, n$. Our task will be to show that the three axioms:

A1. P on X is a suborder,

A2. $DP \subseteq P$,

A3. $PD = P$.

imply that there is a real-valued function u on X such that, for all $x, y \in X$, xPy implies $u(x) > u(y)$. Axiom A2 and the $PD \subseteq P$ part of A3 say that if xPy & yDz or xDy & yPz then xPz. Along with A1 this implies that if xDy then it is false that yPx. The other part of A3, $P \subseteq PD$, says that if xPy then there is a $z \in X$ such that xPz & zDy.

The following six steps outline the proof that {A1, A2, A3} is sufficient for the existence of a real-valued function u on X that gives $u(x) > u(y)$ whenever xPy. The reader is invited to supply the details.

1. Let $P_1 = P \cup D$. P_1 is a suborder.

2. Let $P_2 = P_1^t$, the transitive closure of P_1. Since P_2 is a strict partial order, Szpilrajn's extension theorem assures us that there is a linear order P_3 on X such that $P_2 \subseteq P_3$.

3. Let Y be the set of rational vectors in X. Y is countable. Define a binary relation E on X by

$$xEy \text{ if and only if } x = y \text{ or } (x, y \notin Y \text{ \& } xP_3zP_3y \text{ for no } z \in Y \text{ \& } yP_3zP_3x \text{ for no } z \in Y).$$

E is an equivalence on X.

4. Define P_4 on X/E by

$$aP_4b \text{ if and only if } a \neq b \text{ and } xP_3y \text{ for some } x \in a \text{ \& } y \in b.$$

P_4 on X/E is a linear order. Moreover, $Y' = \{\{x\} : x \in Y\}$ is a subset of X/E and is P_4-order dense in X/E.

5. Thus, by Cantor's Theorem, there is a real-valued function f on X/E such that, for all $a, b \in X/E$, aP_4b if and only if $f(a) > f(b)$.

6. Let $u(x) = f(a)$ when $x \in a$. Suppose xPy. Using A3 to get a z such that xPz & zDy, take $t \in Y$ such that $zDtDy$. Thus $xP_1zP_1tP_1y$ and therefore xP_3tP_3y. Hence not xEy. It follows that, with $x \in a$ & $y \in b$, aP_4b. Therefore $f(a) > f(b)$ and $u(x) > u(y)$.

CHAPTER 8

topology

Our study of order-preserving functions continues under special sets of subsets that are basic to a great deal of mathematics. This chapter concentrates on topologies. The next chapter looks at algebras of sets and probability measures. A basic notion for both chapters is closure under set-theoretic operations.

Let $\mathcal{S}$ be a nonempty set of subsets of a nonempty set X. Then $\mathcal{S}$ is *closed under finite unions* if $A \cup B \in \mathcal{S}$ whenever $A, B \in \mathcal{S}$. If $\mathcal{S}$ is closed under finite unions then $\bigcup_{i=1}^{m} A_i \in \mathcal{S}$ when $A_i \in \mathcal{S}$ for each i since $\cup A_i = ((\ldots ((A_1 \cup A_2) \cup A_3) \ldots) \cup A_m)$. $\mathcal{S}$ is *closed under countable unions* if $\bigcup_{\mathcal{A}} A \in \mathcal{S}$ whenever $\mathcal{A}$ is a nonempty countable subset of $\mathcal{S}$. $\mathcal{S}$ is *closed under unions* if $\bigcup_{\mathcal{A}} A \in \mathcal{S}$ whenever $\phi \subset \mathcal{A} \subseteq \mathcal{S}$. Similar definitions apply for closure under intersections.

In addition, $\mathcal{S}$ is *closed under differences* if $A - B \in \mathcal{S}$ whenever $A, B \in \mathcal{S}$, and is *closed under complementation* if $A^c = X - A$ is in $\mathcal{S}$ whenever $A \in \mathcal{S}$. Algebras of sets are closed under complementation and differences, but this is not generally true of topologies.

Topologies

$\mathcal{S}$ is a *topology* and the pair $(X, \mathcal{S})$ is a *topological space* if

T1. *ϕ and X are in $\mathcal{S}$,*

T2. *$\mathcal{S}$ is closed under finite intersections,*

T3. *$\mathcal{S}$ is closed under unions.*

The elements in a topology $\mathcal{S}$ are called *open sets* or open subsets of X. The complement of each open set is a *closed set*. Since ϕ, $X \in \mathcal{S}$, both ϕ and X are open and closed.

If Y is a nonempty subset of X and $(X, \mathcal{S})$ is a topological space then $(Y, \{A \cap Y : A \in \mathcal{S}\})$ is a topological space also and $\{A \cap Y : A \in \mathcal{S}\}$ is the *relative topology* for Y.

Suppose $(X, \mathcal{S})$ is a topological space. Then $\mathcal{B}$ is a *base* of $\mathcal{S}$ if and only if $\mathcal{B} \subseteq \mathcal{S}$ and each $A \in \mathcal{S}$ equals a union (perhaps an empty union) of elements in $\mathcal{B}$. Thus, ϕ need not be in $\mathcal{B}$ for $\mathcal{B}$ to be a base. It can be shown that $\mathcal{B}$ is a base of $\mathcal{S}$ if and only if $\mathcal{B} \subseteq \mathcal{S}$ and whenever $x \in A$ & $A \in \mathcal{S}$ there is a $B \in \mathcal{B}$ such that $x \in B$ & $B \subseteq A$. $\mathcal{B}$ is a *subbase* of $\mathcal{S}$ if and only if each $A \in \mathcal{S}$ equals a union of finite intersections of elements in $\mathcal{B}$. Thus $\mathcal{B}$ is a subbase if and only if the set of finite intersections of elements in $\mathcal{B}$ is a base.

We shall say that the topology for X *generated* by a nonempty set $\mathcal{A}$ of subsets of X is the topology for X that equals the intersection of all topologies that include $\mathcal{A}$. (It is easily verified that this intersection satisfies T1, T2, and T3). If $\bigcup_{\mathcal{A}} A = X$ then $\mathcal{A}$ is a subbase of the topology for X generated by $\mathcal{A}$. [See, for example, Kelley (1955, pp. 47–48) for this result and others.] In any event $\mathcal{A} \cup \{X\}$ is a subbase of the topology for X that is generated by $\mathcal{A}$.

The topology for $X = Re$ that is generated by the set $\mathcal{B}$ of all half-infinite open intervals $\{x : a < x\}$ and $\{x : x < a\}$ in Re is the *usual topology*. $\mathcal{B}$ is a subbase of the usual topology. The set of all open intervals in Re is a base of the usual topology. With $\mathcal{U}$ the usual topology, $\{U \cap [0, 1] : U \in \mathcal{U}\}$ is the relative usual topology for the closed unit interval $[0, 1]$, which is an open (and closed) set in the relative topology.

Connected and separable spaces

Before introducing an order on a topological space we shall consider several additional definitions.

A topological space $(X, \mathcal{S})$ is *connected* if and only if X cannot be partitioned into two nonempty open sets (in $\mathcal{S}$). The *closure* $\bar{A}$ of any subset $A \subseteq X$ in the space $(X, \mathcal{S})$ is the intersection of all closed subsets of X that include $\bar{A}$. Equivalently,

$$\dot{A} = \{y : y \in X \quad \text{and if} \quad y \in B \ \& \ B \in \mathcal{S} \quad \text{then} \quad A \cap B \neq \phi\}.$$

$\bar{A}$ is a closed subset of X. It is true also that $(X, \mathcal{S})$ is connected if and only if X is not the union of two nonempty subsets A and B for which $\bar{A} \cap B = A \cap \bar{B} = \phi$.

A *subset* Y of X is *connected* if and only if the relative topological space $(Y, \{A \cap Y : A \in \mathcal{S}\})$ is connected. Equivalently $Y \subseteq X$ is connected if and only if for every $A, B \in \mathcal{S}$ it is false that $Y \cap A \neq \phi, Y \cap B \neq \phi, Y \subseteq A \cup B$ and $Y \cap A \cap B = \phi$.

The topological space $(X, \mathcal{S})$ is *separable* if and only if X includes a countable subset whose closure is X. It is easily seen that if $\mathcal{S}$ has a countable base then $(X, \mathcal{S})$ is separable but not necessarily connected. Warning: a space $(X, \mathcal{S})$ for which $\mathcal{S}$ has a countable base is sometimes called separable, as in Luce and Suppes (1965), or perfectly separable, as in Debreu (1954).

The space $(Re, \mathcal{U})$ is connected, separable, and has a countable base, as can be shown using the fact that the set of rational numbers is countable.

Orders on topological spaces

With P a binary relation on X and $(X, \mathcal{S})$ a topological space we shall call $\mathcal{S}$ a

P^+ topology if $\{x : x \in X \;\&\; yPx\} \in \mathcal{S}$ for all $y \in X$;

P^- topology if $\{x : x \in X \;\&\; xPy\} \in \mathcal{S}$ for all $y \in X$;

P topology if it is both a P^+ and P^- topology.

By intersections, $\{x : yPx \;\&\; xPz\} \in \mathcal{S}$ for each $y, z \in X$ when $\mathcal{S}$ is a P topology. If P is a strict partial order then the intersection of all P topologies is sometimes called the P-order topology. It is the topology generated by the preference intervals $\{x : yPx\}$ and $\{x : xPy\}$ for all $y \in X$, and these intervals along with X constitute a subbase for the P-order topology. The $<$-order topology for Re is the usual topology $\mathcal{U}$.

Relational properties are sometimes implied by a combination of other relational properties and topological properties. One of the early results along this line is due to Eilenberg (1941): if $(X, \mathcal{S})$ is connected, $\mathcal{S}$ is a P topology and if P is asymmetric and weakly connected then P is transitive and is therefore a linear order. To prove this we observe first that asymmetry and weak connectedness imply that exactly one of xPy, yPx, and $x = y$ holds for each $x, y \in X$. Suppose that $xPy \;\&\; yPz$. Then $x = z$ is false since it contradicts asymmetry. Hence transitivity holds unless zPx. Suppose in fact that transitivity is false with xPy & yPz & zPx. Letting $A_t = \{s : sPt\}$ and $B_t = \{s : tPs\}$ we then obtain, using the definition of P topology,

$$z \in (A_x \cap B_y) \in \mathcal{S} \text{ and } x \in (A_y \cap B_z) \in \mathcal{S} \text{ and } y \in (A_z \cap B_x) \in \mathcal{S}.$$

By asymmetry, the three two-set intersections are mutually disjoint. Therefore, in order not to violate connectedness of $(X, \mathcal{S})$ there must be elements in no one of $A_x \cap B_y$, $A_y \cap B_z$, and $A_z \cap B_x$. It is easily seen that all such elements must be in $A_x \cap A_y \cap A_z$ or in $B_x \cap B_y \cap B_z$. But then $(A_x \cap B_y) \cup (A_y \cap B_z) \cup (A_z \cap B_x)$ and $(A_x \cap A_y \cap A_z) \cup (B_x \cap B_y \cap B_z)$ are two nonempty disjoint open sets whose union is X, which contradicts connectedness. Hence xPy & yPz & zPx must be false.

Sonnenschein (1965) presents a number of results of this sort. As in the final section of Chapter 3 he takes R (preference-or-indifference) as basic and defines P and I by $P = R \cap R^{c*}$ and $I = R \cap R^*$. One result, due to Rader (1963), is: if R is complete, $\{x : x \in X \text{ \& } xRy\}$ is connected for each $y \in X$, if $PI \subseteq P$ and if $\mathcal{S}$ is a P topology then R is transitive and is therefore a complete preorder. Another result is: if $X \subseteq Re^n$, if xRy and $0 < \alpha < 1$ imply $\alpha x+(1-\alpha)yRy$, if I is transitive and R is complete, and if $\mathcal{S}$ is a P topology then R is transitive.

Real-valued functions

Cantor's Theorem in Chapter 7 gives necessary and sufficient conditions for the existence of a real-valued function u on X that satisfies xPy if and only if $u(x) > u(y)$, for all $x, y \in X$. Various combinations of order and topological properties are also sufficient for the existence of such a u although they are not wholly necessary. For example, there is such a u if $(X, \mathcal{S})$ is a topological space, P on X is a weak order, and either

a. $\mathcal{S}$ has a countable base and $\mathcal{S}$ is either a P^+ topology or a P^- topology, or
b. $(X, \mathcal{S})$ is connected and separable and $\mathcal{S}$ is a P topology.

Result a is due to Rader (1963) and b is due to Eilenberg (1941). Efficient proofs for these and related results in the next section are provided by Debreu (1964). See also Bowen (1968).

Continuity

If $(X, \mathcal{S})$ and $(Y, \mathcal{T})$ are topological spaces and f is a function on X into Y then f is $\mathcal{S}-\mathcal{T}$ *continuous* if and only if $f^{-1}(T) \in \mathcal{S}$ for every $T \in \mathcal{T}$, where $f^{-1}(T) = \{x : x \in X \;\&\; f(x) \in T\}$. If $(Y, \mathcal{T}) = (Re, \mathcal{U})$ and f is $\mathcal{S}-\mathcal{U}$ continuous, we simple say that f is continuous (or continuous in the topology $\mathcal{S}$).

Suppose f is a real-valued function on X with $(X, \mathcal{S})$ a topological space. Then f is *upper semicontinuous* if and only if

$$\{x : x \in X \quad \& \quad f(x) < a\} \in \mathcal{S} \quad \text{for all} \quad a \in Re,$$

and f is *lower semicontinuous* if and only if

$$\{x : x \in X \quad \& \quad f(x) > a\} \in \mathcal{S} \quad \text{for all} \quad a \in Re.$$

Since the half-infinite open intervals are a subbase of the usual topology $\mathcal{U}$, it is easily seen that f is continuous if and only if it is upper semicontinuous and lower semicontinuous.

The general theorem for continuous real-valued functions that satisfy $u(x) > u(y)$ if and only if xPy has been established by Debreu (1964). To use Debreu's term call u *increasing* if and only if, for all $x, y \in X$, $u(x) > u(y)$ if and only if xPy. Suppose that $(X, \mathcal{S})$ is a topological space and there is a real-valued increasing function on X. Then there is a real-valued increasing function on X that

a. is upper semicontinuous whenever $\mathcal{S}$ is a P^+ topology,
b. is lower semicontinuous whenever $\mathcal{S}$ is a P^- topology,
c. is continuous whenever $\mathcal{S}$ is a P topology.

Moreover, these conditions are necessary as well as sufficient for the several forms of continuity. For example, if $\mathcal{S}$ is not a P^+ topology then no increasing function on X can be upper semicontinuous.

These results combine in an obvious way with the conditions for the existence of increasing functions. For example, if $(X, \mathcal{S})$ is a topological space and P is a binary relation on X then there is an upper semicontinuous real-valued function u on X that satisfies

$$xPy \quad \text{if and only if} \quad u(x) > u(y), \quad \text{for all} \quad x, y \in X,$$

if and only if P is a weak order, X includes a countable subset that is P-order dense in X, and $\mathcal{S}$ is a P^+ topology.

Product topologies

Since decision theories often involve products of sets we shall conclude this chapter with a word on product topologies.

When $(X_i, \mathcal{S}_i)$ is a topological space for $i = 1, \ldots, n$ and $X = \Pi X_i$, the *product topology* $\Pi\mathcal{S}_i$ for X is defined in two equivalent ways as follows:

$$\Pi\mathcal{S}_i = \{A : A \subseteq X \text{ and if } (x_1, \ldots, x_n) \in A \text{ then there are } A_i \in \mathcal{S}_i \text{ for which } x_i \in A_i \text{ for all } i \text{ and } \Pi A_i \subseteq A\}$$
$$= \text{the topology whose base is } \{\Pi A_i : A_i \in \mathcal{S}_i \text{ for all } i\}.$$

The usual topology (or usual product topology) for Re^n is denoted $\mathcal{U}^n$, a base for which is $\{\Pi A_i : A_i \in \mathcal{U}$ for all $i\}$. To illustrate continuity in this context let X be the nonnegative orthant of Re^n and let $\mathcal{S} = \{X \cap A : A \in \mathcal{U}^n\}$ be the relative usual topology for X. When axioms W1, W2, and W3 of Chapter 4 hold for P on X, $\mathcal{S}$ is easily shown to be a P topology. Hence, by Debreu's continuity theory there is a real-valued increasing function u on X that is continuous (in the topology $\mathcal{S}$).

With $(X, \mathcal{S})$ as just defined, A1, A2 and A3 in the final section of Chapter 7 imply that there is an upper semicontinuous real-valued function v on X such that, for all $x, y \in X$, xPy implies $v(x) > v(y)$. The proof of upper semicontinuity for this case can be based on the following proposition (which you should not find too difficult to prove): if $(X, \mathcal{S})$ is a topological space and u is a bounded real-valued function on X then v on X, defined by

$$v(x) = \inf\{\sup\{u(y) : y \in A\} : A \in \mathcal{S} \text{ and } A \text{ contains } x\},$$

is upper semicontinuous (in $\mathcal{S}$). This definition of v is usually abbreviated as $v = \lim\sup u$.

A basic result for additivity with product spaces has been established by Debreu (1960). We state his theorem for $n \geqslant 3$: the $n = 2$ version of the theorem is Theorem 5.4 in Fishburn (1970). Suppose $n \geqslant 3$, $X = \Pi X_i$, P on X is a weak order, xPy for some $x, y \in X$ that differ only in the ith components ($i = 1, \ldots, n$), and the following hold throughout X:

D1. *If* $\{x_i, z_i\} = \{y_i, w_i\}$ *for* $i = 1, \ldots, n$ *and if* xPy *or* $x(P \cup P^*)^c y$ *then not* zPw.

D2. *Each* $(X_i, \mathcal{S}_i)$ *is a connected and separable topological space.*

D3. $\Pi\mathcal{S}_i$ *is a* P *topology.*

Then for each i there is a continuous (in $\mathcal{S}_i$) real-valued function u_i on X_i such that, for all $x, y \in X$,

$$xPy \quad \text{if and only if} \quad \sum_{i=1}^{n} u_i(x_i) > \sum_{i=1}^{n} u_i(y_i),$$

and if $v_1, \ldots, v_n$ satisfy this in place of $u_1, \ldots, u_n$ then there are numbers $a, b_1, \ldots, b_n$ with $a > 0$ such that, for each i, $v_i(x_i) = au_i(x_i)+b_i$ for all $x_i \in X_i$.

Variations of this theorem and the $n = 2$ version for related areas of preference theory are discussed by Fishburn (1970: Section 6.3, Theorems 7.4 and 7.5, Exercises 13.16–13.19).

The reader who is interested in pursuing the relationship between order and topology should consult the book by Nachbin (1965) as well as the references cited above. Pfanzagl (1968) also presents a variety of theory that involves topological concepts. We shall consider some of this in Chapter 11.

CHAPTER 9

probability

This chapter examines several types of probability measures defined on algebras of subsets of a set X, discusses axiomatic bases for such measures, and presents a brief outline of the theory of mathematical expectation. The concepts that are defined in this chapter will be used extensively in Chapter 10.

Algebras of subsets

As in the preceding chapter we shall assume that $\mathcal{S}$ is a nonempty set of subsets of a nonempty set X. $\mathcal{S}$ is an *algebra of subsets* of X if and only if

A1. *$\mathcal{S}$ is closed under finite unions,*

A2. *$\mathcal{S}$ is closed under complementation.*

A2 says that $A^c = X - A$ is in $\mathcal{S}$ when $A \in \mathcal{S}$. When $\mathcal{S}$ is an algebra of subsets, the elements in $\mathcal{S}$ are said to be *$\mathcal{S}$-measurable* subsets of X.

Because $\mathcal{S} \neq \phi$, an algebra contains X (or X is $\mathcal{S}$-measurable) since $X = A \cup A^c$. Since $A \cap B = (A^c \cup B^c)^c$, an algebra is closed under finite intersections. That is, if $A, B \in \mathcal{S}$ then $A \cap B \in \mathcal{S}$. Since $A - B = A \cap B^c = (A^c \cup B)^c$, an algebra is closed under differences.

An algebra of subsets $\mathcal{S}$ is a *σ-algebra* (or Borel algebra) if $\mathcal{S}$ is closed under countable unions. A σ-algebra $\mathcal{S}$ is closed under countable intersections since if $\mathcal{A}$ is a nonempty countable subset of $\mathcal{S}$ then $\bigcap_{\mathcal{A}} A \in \mathcal{S}$ since $\bigcap_{\mathcal{A}} A = (\bigcup_{\mathcal{A}} A^c)^c$. If X is denumerable and $\mathcal{S} = \{A\colon$ either A or A^c is a finite subset of $X\}$ then $\mathcal{S}$ is an algebra but not a σ-algebra. (Why is this so?)

Sometimes an $\mathcal{S}$ is both an algebra and a topology. This is true of $\mathcal{S} = \{\phi, X\}$ and of the set of all subsets of X. An algebra that is not a σ-algebra cannot be a topology. In general, topologies need not be closed under complementation (A2), and algebras need not be closed under unions (T3). If $\phi \subset A \subset X$ then $\{\phi, A, X\}$ is a topology but not an algebra.

If $\mathcal{A}$ is a nonempty set of subsets of X then the algebra (σ-algebra) *generated by* $\mathcal{A}$ is the intersection of all algebras (σ-algebras) that include $\mathcal{A}$. Unlike the case with topologies, there is in general no simple constructive process using sets in $\mathcal{A}$ that describes the algebra generated by $\mathcal{A}$. (Recall that the topology generated by $\mathcal{A}$ consists of those subsets of X that are unions of finite intersections of elements in $\mathcal{A} \cup \{X\}$.)

An algebra $\mathcal{S}$ is *finite* or *simple* if $\mathcal{S}$ is finite. When $\mathcal{S}$ is finite, there is a unique partition $\mathcal{A}$ of X such that $\mathcal{A} \subseteq \mathcal{S}$ and every $B \in \mathcal{S}$ can be written as a finite union (perhaps an empty union to include ϕ) of elements in $\mathcal{A}$. $\mathcal{S}$ of course is generated by $\mathcal{A}$ in the simple case.

Probability measures

Let $\mathcal{S}$ be an algebra of subsets of X. A *probability measure* on $\mathcal{S}$ is a real-valued function p on $\mathcal{S}$ that satisfies

B1. $p(A) \geqslant 0$ for all $A \in \mathcal{S}$,

B2. $p(X) = 1$,

B3. $p(A \cup B) = p(A)+p(B)$ whenever $A, B \in \mathcal{S}$ and $A \cap B = \phi$.

Property B3 says that p is finitely additive: if $\mathcal{A}$ is a nonempty finite set of mutually disjoint elements in $\mathcal{S}$ then $p(\bigcup_{\mathcal{A}} A) = \sum_{\mathcal{A}} p(A)$.

Since $A \cup A^c = X$ and $A \cap A^c = \phi$, B3 and B2 yield $p(A)+p(A^c) = 1$ for every $A \in \mathcal{S}$. Also, since $A = (A \cap B) \cup (A \cap B^c)$, $B = (A \cap B) \cup (A^c \cap B)$ and $A \cup B = (A \cap B) \cup (A \cap B^c) \cup (A^c \cap B)$, it follows from finite additivity that

$$\begin{aligned} p(A) &= p(A \cap B)+p(A \cap B^c) \\ p(B) &= p(A \cap B)+p(A^c \cap B) \\ p(A \cup B) &= p(A \cap B)+p(A \cap B^c)+p(A^c \cap B) \end{aligned}$$

so that $p(A \cup B) = p(A)+p(B)-p(A \cap B)$ whenever $A, B \in \mathcal{S}$.

We shall call a probability measure p on $\mathcal{S}$ *simple* if and only if there is a finite $A \in \mathcal{S}$ for which $p(A) = 1$. If $\{x\} \in \mathcal{S}$ for every $x \in X$ and if p is simple then $p(x) = 0$ for all but a finite number of $x \in X$ and, for all $A \in \mathcal{S}$, $p(A) = \sum_A p(x)$.

A probability measure p on an algebra $\mathcal{S}$ is *countably additive* if and only if

$$p(\bigcup_{\mathcal{A}} A) = \sup\{\textstyle\sum_{\mathcal{B}} p(A) : \mathcal{B} \quad \text{is a finite subset of} \quad \mathcal{A}\}$$

whenever $\mathcal{A}$ is a nonempty countable subset of $\mathcal{S}$ whose elements are mutually disjoint and $\bigcup_{\mathcal{A}} A \in \mathcal{S}$. This definition applies whether or not $\mathcal{S}$ is a σ-algebra. However, σ-algebras are normally used with countably-additive measures, as in Loève (1963). Every simple measure is countably additive. If X is denumerable then a measure p that satisfies B1 through B5 in the final section of Chapter 6 is not countably additive since $\sum_B p(x) = 0$ for any finite $B \subseteq X$ and hence $\sup\{\ldots\} = 0$, whereas $p(X) = 1$. (In this example take $\mathcal{A} = \{\{x\} : x \in X\}$.)

The following so-called continuity condition is an important property of countably-additive probability measures that is not generally true for measures that are only finitely additive:

if $\mathcal{A}$ is a nonempty countable subset of $\mathcal{S}$,

if $\subset$ on $\mathcal{A}$ is a linear order, and

if $\bigcup_{\mathcal{A}} A \in \mathcal{S}$,

then $p(\bigcup_{\mathcal{A}} A) = \sup\{p(A) : A \in \mathcal{A}\}$.

Discrete probability measures are special kinds of countably-additive measures. p on $\mathcal{S}$ is *discrete* if and only if $\{x\} \in \mathcal{S}$ for every $x \in X$, $\mathcal{S}$ is a σ-algebra, p is countably additive and $p(A) = 1$ for some countable $A \in \mathcal{S}$. If p on $\mathcal{S}$ is discrete then $p(x) = 0$ for all but a countable number of $x \in X$ and

$$p(A) = \textstyle\sum_A p(x) = \sup\{\sum_B p(x) : B \quad \text{is a finite subset of} \quad A\}.$$

Discrete measures are used extensively by Blackwell and Girshick (1954).

If p is a probability measure on $\mathcal{S}$ and if $A \in \mathcal{S}$ and $p(A) > 0$ then p_A defined by

$$p_A(B) = p(B \cap A)/p(A) \qquad \text{for all} \qquad B \in \mathcal{S}$$

is a probability measure on $\mathcal{S}$ and is called the *conditional measure* of p given A. If $A, B \in \mathcal{S}$ and $p(A)p(B) > 0$ then

$$p(A)p_A(B) = p(B)p_B(A) = p(A \cap B),$$

and $p_B(A) = p(A)p_A(B)/p(B)$, which is sometimes referred to as Bayes' rule.

Axioms for probability

Just as we have given preference axioms that give rise to order-preserving utility functions, it is possible to axiomatize a qualitative probability relation on an algebra $\mathcal{S}$ that gives rise to an order-preserving probability measure. The main difference between the mathematics used in the two cases involves the additivity property B3 for probability measures.

We can interpret X as a set of events, one and only one of which will occur in a given situation. The decision maker is uncertain of which will occur and therefore we consider quantifying his uncertainty by a probability measure.

Suppose that $\mathcal{S}$ is a simple algebra and $\mathcal{A}$ is the unique partition of X that generates $\mathcal{S}$. With $\mathcal{A} = \{A_1, \ldots, A_n\}$ and $B_1, \ldots, B_m, C_1, \ldots, C_m \in \mathcal{S}$ let $\sum_{j=1}^m B_j = \sum_{j=1}^m C_j$ mean that, for each A_i, the number of B_j that include A_i equals the number of C_j that include A_i. With APB interpreted as 'A is more probable than B' and $I = (P \cup P^*)^c$ suppose that the following axioms hold for all $A_i \in \mathcal{A}$, all $B_1, \ldots, B_m, C_1, \ldots, C_m \in \mathcal{S}$ and all $m \geq 2$:

1. *not* $\phi P A_i$,
2. $XP\phi$,
3. *If* $\sum_{j=1}^m B_j = \sum_{j=1}^m C_j$ *and if* $B_j(P \cup I)C_j$ *for each* $j < m$ *then it is false that* $B_m P C_m$.

Then, using Lemma 1 in Chapter 5, it follows that there is a probability measure p on $\mathcal{S}$ such that, for all $A, B \in \mathcal{S}$, APB if and only if $p(A) > p(B)$. This result is due to Kraft, Pratt, and Seidenberg (1959) and Scott (1964). Several weaker versions of these axioms are discussed by Fishburn (1969).

Another case, due to Savage (1954), takes $\mathcal{S}$ as the set of all subsets of X and assumes that, for all $A, B, C \in \mathcal{S}$:

1. *not* ϕPA,
2. $XP\phi$,
3. P *on* $\mathcal{S}$ *is a weak order*,
4. *If* $A \cap C = B \cap C = \phi$ *then* APB *if and only if* $(A \cup C)P(B \cup C)$,
5. *If* APB *then there is a finite partition* $\mathcal{C}$ *of* X *such that* $AP(B \cup C)$ *for every* $C \in \mathcal{C}$.

As shown by Savage (1954) or Fishburn (1970, Section 14.2) these axioms imply that there is a unique probability measure p on $\mathcal{S}$ such that APB if and only if $p(A) > p(B)$, and for this p it is true that whenever $B \in \mathcal{S}$ and $0 \leqslant \alpha \leqslant 1$ there is some $C \subseteq B$ for which $p(C) = \alpha p(B)$. The first four axioms in this case identify P as a *qualitative probability* as discussed by de Finetti (1937). The qualitative probability axioms 1 through 4 are necessary but not sufficient for the APB if and only if $p(A) > p(B)$ conclusion. The fifth axiom is a rather powerful Archimedean axiom as is seen from the $p(C) = \alpha p(B)$ property.

Other studies that discuss qualitative axioms for probability include Koopman (1940, 1941), Halphen (1955), Villegas (1964, 1967), Luce (1967, 1968), Domotor (1969), and the papers in Kyburg and Smokler (1964).

Mathematical expectation

Suppose p is a simple probability measure on $\mathcal{S}$, $\{x\} \in \mathcal{S}$ for every $x \in X$, and f is a real-valued function on X. Then $E(f, p)$, the expected value of f with respect to p, is defined by

$$E(f, p) = \textstyle\sum_X f(x)p(x).$$

Suppose p specifies the probabilities of 'heads' and 'tails' on one flip of a coin. Then, if $f(\text{heads}) = \$0$ and $f(\text{tails}) = \$100$,

$$E(f, p) = \$0p(\text{heads}) + \$100p(\text{tails})$$

which gives the expected monetary winnings for gamble f.

In general, $E(f, p)$ will be defined only for real-valued functions f on X that are $\mathcal{S}$-measurable. We say that f is *$\mathcal{S}$-measurable* if and only if $\{x : f(x) \in I\} \in \mathcal{S}$ for every interval I of real numbers. Since $[a, a]$ is an interval when a is a number, each subset of X on which f is constant must be in $\mathcal{S}$ when f is $\mathcal{S}$-measurable. If $\mathcal{S}$ is the set of all subsets of X then every real-valued function f on X is $\mathcal{S}$-measurable.

An $\mathcal{S}$-measurable function f on X is *simple* if and only if $\{f(x): x \in X\}$ is finite. If f is simple and takes on n distinct values $c_1, \ldots, c_n$ with $f(x) = c_i$ if and only if $x \in A_i$, we define

$$E(f, p) = \textstyle\sum_{i=1}^{n} c_i p(A_i).$$

An $\mathcal{S}$-measurable function f on X is *bounded* if and only if there are numbers a and b such that $a \leqslant f(x) \leqslant b$ for all $x \in X$. In this case we define

$$E(f, p) = \sup\{E(f_n, p) : n = 1, 2, \ldots\}$$

where $f_1, f_2, \ldots$ is any sequence of simple $\mathcal{S}$-measurable functions that *converges uniformly from below to f*, which means that

1. $f_1(x) \leqslant f_2(x) \leqslant \ldots$ for all $x \in X$,
2. $f(x) = \sup\{f_n(x): n = 1, 2, \ldots\}$ for all $x \in X$,
3. For any $\delta > 0$ there is a positive integer n (which can depend on δ) such that $f(x) \leqslant f_n(x) + \delta$ for all $x \in X$.

Such a sequence of simple $\mathcal{S}$-measurable functions exists for any bounded $\mathcal{S}$-measurable f as is seen by defining

$$A_1^n = \{x: a \leqslant f(x) \leqslant a+(b-a)/n\}$$
$$A_i^n = \{x: a+(i-1)(b-a)/n < f(x) \leqslant a+i(b-a)/n\} \quad i = 2, \ldots, n$$

and

$$f_n(x) = a+(i-1)(b-a)/n \text{ for all } x \in A_i^n \ (i = 1, \ldots, n).$$

It is not difficult to show that if $f_1, f_2, \ldots$ and $g_1, g_2, \ldots$ are two sequences of simple $\mathcal{S}$-measurable functions that converge uniformly from below to bounded f then $\sup\{E(f_n, p): n = 1, 2, \ldots\} = \sup\{E(g_n, p): n = 1, 2, \ldots\}$. If $\mathcal{S}$ is a σ-algebra and p is countably additivity then uniformity (3) is not needed in defining $E(f, p)$, but (3) may be needed in other cases. For example, $E(f, p) = 1$ if $X = \{0, 1, 2, \ldots\}$, $\mathcal{S}$ is the set of all subsets of X, and $f(x) = x/(1+x)$ and $p(x) = 0$ for all $x \in X$. However, it is easy to construct a sequence $g_1, g_2, \ldots$ of simple functions that converges nonuniformly from below to f [for which (1) and (2) hold and (3) fails] and has $\sup\{E(f_n, p)\colon n = 1, 2, \ldots\} = 0$.

Given any $\mathcal{S}$-measurable function f on X, bounded or unbounded, define f^+ and f^- on X by

$$f^+(x) = f(x) \quad \text{and} \quad f^-(x) = 0 \qquad \text{for all} \quad x \in \{x\colon f(x) \geqslant 0\}$$

$$f^+(x) = 0 \qquad \text{and} \quad f^-(x) = f(x) \quad \text{for all} \quad x \in \{x\colon f(x) < 0\}.$$

We then define $E(f^+, p)$ from bounded functions by

$$E(f^+, p) = \sup\{E[\inf\{f^+(x), c\}, p]\colon c \geqslant 0\}$$

where $E(f^+, p) = \infty$ if $\{E[\inf\{f^+(x), c\}, p]\colon c \geqslant 0\}$ is unbounded above. Similarly, $E(f^-, p) = -\sup\{E[\inf\{-f^-(x), c\}, p]\colon c \geqslant 0\}$, with $E(f^-, p) = -\infty$ in the unbounded case. Then, with $\infty + \text{number} = \infty$ and $\text{number} + (-\infty) = -\infty$, we define $E(f, p)$ by $E(f, p) = E(f^+, p) + E(f^-, p)$ except when $E(f^+, p) = \infty$ and $E(f^-, p) = -\infty$, in which case $E(f, p)$ is not well defined.

For bounded $\mathcal{S}$-measurable functions f and g it is not too difficult to verify that $E(f+g, p) = E(f, p) + E(g, p)$, that $E(cf, p) = cE(f, p)$ when $c \in Re$, and that if $p(A) = 1$ and $f(x) \geqslant g(x)$ for all $x \in A$ then $E(f, p) \geqslant E(g, p)$. For unbounded functions it is possible to have $E(f+g, p)$ well defined when neither $E(f, p)$ nor $E(g, p)$ is well defined: consider, for example, $g = -f$ so that $f + g \equiv 0$.

If p and q are probability measures on $\mathcal{S}$ and $0 \leqslant \alpha \leqslant 1$ then $\alpha p + (1-\alpha)q$ is the probability measure on $\mathcal{S}$ that assigns probability $\alpha p(A) + (1-\alpha)q(A)$ to each $A \in \mathcal{S}$. If f is bounded and $\mathcal{S}$-measurable then $E(f, \alpha p + (1-\alpha)q) = \alpha E(f, p) + (1-\alpha)E(f, q)$. It follows from this that if f is bounded and $\mathcal{S}$-measurable and if $\{A_1, \ldots, A_n\}$ is an $\mathcal{S}$-partition of X (so that $A_i \in \mathcal{S}$ for each i) then

$$E(f, p) = \textstyle\sum_{\{i\colon p(A_i) > 0\}} p(A_i) E(f, p_{A_i}).$$

If p is countably additive, this holds also for denumerable partitions.

Probability from preference

Earlier in this chapter we considered qualitative axioms for probability that give rise to order-preserving probability measures. It is possible also to define a qualitative probability relation from a preference relation as is done by Savage (1954). We shall conclude this chapter with a somewhat different approach than that used by Savage in order to bring expectations into the picture.

Formally, we shall let $\mathcal{S}$ be the set of all subsets of a nonempty set X and let F be the set of all functions on X into [0, 1]. We shall interpret a binary relation P on F as a preference relation. To motivate the axioms that follow we might suppose that if the decision maker 'chooses' $f \in F$ and if $x \in X$ occurs or obtains then he will win a prize of £1000 with probability $f(x)$ and win nothing with probability $1-f(x)$. Thus if $f(x) = \alpha$ and $g(x) = \beta$ for all $x \in X$ and if $\alpha > \beta$ then we would expect that fPg. In the axioms $\alpha f+(1-\alpha)g$ is the function in F that takes the value $\alpha f(x)+(1-\alpha)g(x)$ for each $x \in X$. Event $A \in \mathcal{S}$ is *null* if and only if fIg whenever $f(x) = g(x)$ for every $x \in A^c$. Here and in axiom 6, $I = (P \cup P^*)^c$. We define αPg (and $fP\alpha$) to mean that hPg (fPh) whenever $h(x) = \alpha$ for all $x \in X$. Our axioms are:

1. *P on F is a weak order,*
2. *fPg when $f \equiv 1$ and $g \equiv 0$,*
3. *If fPg and $0 < \alpha < 1$ then $\alpha f+(1-\alpha)hP\alpha g+(1-\alpha)h$,*
4. *If A is a nonnull event, if $f(x) = \alpha$ and $g(x) = \beta$ for all $x \in A$, and if $f(x) = g(x)$ for all $x \in A^c$, then fPg if and only if $\alpha > \beta$,*
5. *If $fPgPh$ then $\alpha f+(1-\alpha)hPgP\beta f+(1-\beta)h$ for some $\alpha, \beta \in]0, 1[$,*
6. *If $f(x)Pg$ for all $x \in X$ or if $fPg(x)$ for all $x \in X$ then $f(P \cup I)g$.*

It follows from Theorem 13.3 in Fishburn (1970) that these axioms imply that there is one and only one probability measure p on $\mathcal{S}$ such that

$$fPg \quad \text{if and only if} \quad E(f, p) > E(g, p), \quad \text{for all} \quad f, g \in F,$$

with $p(A) = 0$ if and only if A is a null event. Moreover, as you can easily see, the axioms are also necessary for the existence of such a p.

In the foregoing order-preserving expression we can interpret p as the decision maker's personal probability measure on $\mathcal{S}$, with $p(A)$ representing his degree of belief in the truth of the proposition 'event A will occur or obtain'. $E(f, p)$ represents his 'total probability' that he will win the prize of £1000 if he 'chooses' f. If we define $f_A \in F$ by $f_A(x) = 1$ if $x \in A$ and $f_A(x) = 0$ if $x \in A^c$, then a binary relation P' on $\mathcal{S}$ defined from P on F by

$$AP'B \quad \text{if and only if} \quad f_A P f_B$$

can be viewed as a qualitative probability relation. $f_A P f_B$ means that he would rather take gamble f_A (win £1000 if A obtains) than gamble f_B (win £1000 if B obtains). Hence it seems reasonable to say that he regards A as more probable than B when $f_A P f_B$. With P' defined as indicated we then have

$$AP'B \quad \text{if and only if} \quad p(A) > p(B)$$

for all $A, B \in \mathcal{S}$ since $E(f_A, p) = p(A)$ for all $A \in \mathcal{S}$.

CHAPTER 10

mixture sets

Mixture sets have been used primarily in decision theories that involve probability and expectation. Our definition of such a set is taken from Herstein and Milnor (1953). After presenting a basic theorem for mixture sets that stems from the von Neumann and Morgenstern (1947) expected-utility theory we shall consider some special theories based on mixture sets.

Mixture sets

A *mixture set* is a nonempty set F and a function that assigns an element $\alpha f+(1-\alpha)g$ in F to each $\alpha \in [0, 1]$ and $(f, g) \in F \times F$ such that, for all $f, g \in F$ and $\alpha, \beta \in [0, 1]$:

M1. $1f+0g=f$,
M2. $\alpha f+(1-\alpha)g=(1-\alpha)g+\alpha f$,
M3. $\alpha[\beta f+(1-\beta)g]+(1-\alpha)g=\alpha\beta f+(1-\alpha\beta)g$.

Other properties follow from M1–M3, including

M4. $\alpha f+(1-\alpha)f=f$,
M5. $\alpha[\beta f+(1-\beta)g]+(1-\alpha)[\gamma f+(1-\gamma)g]=[\alpha\beta+(1-\alpha)\gamma]f+[\alpha(1-\beta)+(1-\alpha)(1-\gamma)]g$.

Luce and Suppes (1965, p. 288) give an explicit proof of M5.

Any nonempty convex subset of Re^n is a mixture set when $\alpha x+(1-\alpha)y$ is interpreted in the usual way. The set F in the last section of Chapter 9 is a mixture set. Any nonempty set of probability measures, each of which is defined on the same algebra, is a mixture set when it is closed under finite convex combinations so that $\alpha p+(1-\alpha)q$ is in the set when p and q are in the set and $0 \leqslant \alpha \leqslant 1$. $\alpha p+(1-\alpha)q$ is the measure that assigns probability $\alpha p(A)+(1-\alpha)q(A)$ to each A in the algebra.

Let P, interpreted as preference, be a binary relation on a mixture set F. We consider the following proposition (after von Neumann and Morgenstern):

NM. *There is a real-valued function u on F such that, for all $f, g \in F$ and $\alpha \in [0, 1]$,*

$$fPg \quad \textit{if and only if} \quad u(f) > u(g),$$
$$u(\alpha f+(1-\alpha)g) = \alpha u(f)+(1-\alpha)u(g).$$

The latter property is referred to as linearity (u is a linear function) or as the expectation property. It is not difficult to prove that if u satisfies the *NM* proposition then a real-valued function v on F satisfies the proposition in place of u if and only if there are numbers $a > 0$ and b such that $v(f) = au(f)+b$ for all $f \in F$. In this case we say that u is unique up to a positive ($a > 0$) linear transformation. If u satisfies the *NM* proposition and v is a monotonic transformation of u, so that $v(f) > v(g)$ if and only of $u(f) > u(g)$ then v preserves order but is linear only if the transformation is linear.

Several sets of necessary and sufficient axioms for the *NM* proposition have been developed. Two of the most interesting of these are those given by Herstein and Milnor (1953) and by Jensen (1967). The latter set consists of the following three axioms: for all $f, g, h \in F$,

A1. *P on F is a weak order,*
A2. *If fPg and $0 < \alpha < 1$ then $\alpha f+(1-\alpha)hP\alpha g+(1-\alpha)h$,*
A3. *If fPg and gPh then there are $\alpha, \beta \in]0, 1[$ such that $\alpha f+(1-\alpha)hPg$ and $gP\beta f+(1-\beta)h$.*

A2 is a form of independence axiom and A3 is an Archimedean property. These axioms are discussed also by Fishburn (1970, Chapter 8), where a detailed proof of the *NM* proposition is given. Among other things, the proof involves a demonstration, due to Jensen, that A1–A3 imply that if fIg and $0 < \alpha < 1$ then $\alpha f+(1-\alpha)hI\alpha g+(1-\alpha)h$, where $I = (P \cup P^*)^c$. Then it is shown that, if $f(P \cup I)g(P \cup I)h$ and fPh, there is a unique $\alpha \in [0, 1]$ for which $gI\alpha f+(1-\alpha)h$. The construction of u follows from this property. For example, suppose that F has a most preferred element f^* and a least preferred element f_*, with f^*Pf_*. If we then set $u(f) = \alpha$ when $fI\alpha f^*+(1-\alpha)f_*$, u can be shown to satisfy the *NM* proposition.

The Herstein–Milnor axioms are: for all $f, g, h \in F$,

B1. *P on F is a weak order,*

B2. *If* fIg *then* $\frac{1}{2}f+\frac{1}{2}hI\frac{1}{2}g+\frac{1}{2}h$,

B3. $\{\alpha : \alpha \in [0, 1]$ *and* $hP\alpha f+(1-\alpha)g\}$ *and* $\{\alpha : \alpha \in [0, 1]$ *and* $\alpha f+(1-\alpha)gPh\}$ *are open sets in the relative usual topology for* $[0, 1]$.

Luce and Suppes (1965) also prove the *NM* proposition on the basis of these axioms. Their proof is similar to the Herstein–Milnor proof. Since each of {A1, A2, A3} and {B1, B2, B3} is necessary and sufficient for the *NM* proposition, each set of axioms implies the axioms in the other set.

Expected utility

The setting for the expected utility of von Neumann and Morgenstern (1947) arose in a study of the theory of games. Each player in a game has a set of strategies, and the adoption of a strategy by each player leads to a payoff or joint consequence for the players. Each player will evaluate the possible payoffs according to his own preferences. With very little extra expense a player may decide to use a mixed strategy, which means that he will select a strategy according to the outcome of a chance experiment (like flipping a coin) that assigns specified probabilities to his basic or pure strategies. When this is done and a specified strategy is assumed for each of the other players (including possible mixed strategies for others), the resultant payoff will not be uniquely specified beforehand but rather will be determined according to a probability measure that arises from the assumed mixed strategies.

In order to quantify individuals' preferences under these conditions we can consider the set of all simple probability measures on the set of all subsets of consequences and, for each individual, suppose that his preferences satisfy A1, A2 and A3 on this mixture set. If $\mathcal{P}$ is the set of simple measures and P is the preference relation for a given player, it follows easily from the *NM* proposition that there is a real-valued function u on the set X of consequences such that, for all $p, q \in \mathcal{P}$,

$$pPq \quad \text{if and only if} \quad E(u, p) > E(u, q).$$

u on X is defined from u on $\mathcal{P}$ by setting $u(x) = u(p)$ when $p(x) = 1$. The displayed expected-utility representation incorporates both the order-preserving property and the linearity property of the *NM* proposition.

More generally, we can consider other types of sets of probability measures and, using the *NM* proposition as a beginning point, obtain expected-utility representations for these sets. This is done by Blackwell and Girshick (1954) for discrete measures and by Fishburn (1970, Chapter 10) for several other sets of measures.

In the latter theory it is assumed that X is a nonempty set of consequences, that $\mathcal{S}$ is an algebra of subsets of X that contains each $\{x\}$ for $x \in X$ and all sets of the form $\{x: xPy\}$ and $\{x: yPx\}$ for each $y \in X$, and that $\mathcal{P}$ is a set of probability measures on $\mathcal{S}$ that (1) contains each measure that assigns probability 1 to a unit subset of X, (2) is closed under countable convex combinations so that $\sum_{i=1}^{\infty} \alpha_i p^i \in \mathcal{P}$ when $\alpha_i > 0$ and $p^i \in \mathcal{P}$ for $i = 1, 2, \ldots$ and $\sum \alpha_i = 1$, and (3) is closed under the formation of conditional probabilities so that $p_A \in \mathcal{P}$ whenever $p \in \mathcal{P}$, $A \in \mathcal{S}$ and $p(A) > 0$. Condition (1) for $\mathcal{P}$ enables us to define u on X from u on $\mathcal{P}$ in an unambiguous way and the conditions for $\mathcal{S}$ assure that u on X is $\mathcal{S}$-measurable so that $E(u, p)$ can be defined for each $p \in \mathcal{P}$. The two closure conditions for $\mathcal{P}$ permit the conclusion that, for each $p \in \mathcal{P}$, $E(u, p) = u(p)$, where u in $E(u, p)$ is on X and $u(p)$ is as given in the *NM* proposition.

Under the special conditions for $\mathcal{S}$ and $\mathcal{P}$, Fishburn's theory uses A1, A2, and A3 for P on the mixture set $\mathcal{P}$ along with one other axiom, as follows, in which pPx (or xPp) means that pPp' (or $p'Pp$) when $p'(x) = 1$:

A4. *If* $A \in \mathcal{S}$, $q(A) = 1$, *and* pPx *for all* $x \in A$ *then not* qPp.

If $A \in \mathcal{S}$, $q(A) = 1$, *and* xPp *for all* $x \in A$ *then not* pPq.

The first part of A4 says that if you prefer the measure p as a whole to each consequence in an $\mathcal{S}$-measurable subset A of X for which $q(A) = 1$ then you will not prefer q to p. The second part of A4 is the symmetric dual of the first part.

From the structural conditions on $\mathcal{S}$ and $\mathcal{P}$ along with A1–A4 it follows that there is a *bounded* real-valued function u on X that satisfies pPq if and only if $E(u, p) > E(u, q)$, for all $p, q \in \mathcal{P}$. If every $p \in \mathcal{P}$ is countably additive then the same conclusion follows when A4 is replaced by

A4′. *If* $A \in \mathcal{S}$, $q(A) = 1$, *and* $y(P \cup I)x$ *for all* $x \in A$ *then* $y(P \cup I)q$, and its symmetric dual. If every $p \in \mathcal{P}$ is discrete and if xPy for some $x, y \in X$, the same conclusion follows when A4 is replaced by

A4″. *If* $A \in \mathcal{S}$, $q(A) = 1$, *and* yPx *for all* $x \in A$ *then not* qPy, and its symmetric dual.

Strict partial orders

There are several mixture-set axiomatizations with applications for expected utility that do not assume all of A1, A2, and A3. Since these three axioms are necessary and sufficient for the *NM* proposition, the axioms that we consider in this section and the next do not imply that proposition.

In Chapter 5 (using somewhat different notation) we considered axioms that give pPq implies $E(u, p) > E(u, q)$, for all p, q in some finite set of simple probability measures (which is not generally a mixture set). Following Aumann (1962), it is possible to extend this result to infinite sets of measures provided that this set is a mixture set and the algebra $\mathcal{S}$ on which the measures are defined is simple and contains each $\{x\}$. To illustrate this let X be finite, let $\mathcal{S}$ be the set of all subsets of X, let $\mathcal{P}$ be the set of all probability measures on $\mathcal{S}$, and suppose that the following hold for all $p, q, r, s \in \mathcal{P}$:

C1. *P on $\mathcal{P}$ is a strict partial order,*
C2. *If $0 < \alpha < 1$ then pPq if and only if $\alpha p+(1-\alpha) rP\alpha q+(1-\alpha) r$,*
C3. *If $\alpha p+(1-\alpha) rP\alpha q+(1-\alpha) s$ for all $0 < \alpha \leq 1$ then not sPr.*

Then there is a real-valued function u on X such that, for all $p, q \in \mathcal{P}$,

$$pPq \quad \text{implies} \quad E(u, p) > E(u, q).$$

The proof of this, following Aumann (1962) or Fishburn (1970, Chapter 9), is based on a theorem for convex cones in Re^n. $C \subseteq Re^n$ is a *cone* if and only if $\alpha x \in C$ whenever $x \in C$ and $\alpha > 0$. C is a convex cone if and only if $\alpha x+\beta y \in C$ whenever $x, y \in C$ and $\alpha, \beta > 0$. Aumann's convex cone theorem is: *if C is a nonempty convex cone in Re^n and if $\bar{C} \cup (-C) = \phi$* (where $\bar{C}$ is the closure of C in the usual topology and $-C = \{x\colon -x \in C\}$) *then there is a $\varrho \in Re^n$ such that*

$$\varrho \cdot x > 0 \quad \textit{for all} \quad x \in C.$$

This is a generalization of Lemma 1, Chapter 5, with $M = K$. The failure of $\bar{C} \cap (-C) = \phi$ even when $C \cap (-C) = \phi$ can be seen from Figure 2-II when Z there is extended to the smallest cone in Re^2 that includes Z. The connection between the cone theorem and $E(u, p) > E(u, q)$ is easily seen when finite X is exhaustively enumerated as $\{x_1, \ldots, x_n\}$ and $E(u, p)$ is written as $u(x_1) p(x_1)+ \ldots +u(x_n) p(x_n)$, so that $E(u, p) > E(u, q)$ if and only if $u(x_1)[p(x_1)-q(x_1)]+\ldots+u(x_n)[p(x_n)-q(x_n)] > 0$. We take C as the convex cone generated by all vectors $(p(x_1)-q(x_1), \ldots, p(x_n)-q(x_n))$ for which pPq and identify ϱ with $(u(x_1), \ldots, u(x_n))$. The proof is carried through by showing that C1, C2 and C3 imply that $\bar{C} \cap (-C) = \phi$.

Lexicographic NM proposition

An earlier development of Hausner (1954) generalizes the *NM* proposition to a lexicographically ordered form. With $(a_1, \ldots, a_n) \overset{L}{>} (b_1, \ldots, b_n)$ if and only if $a \neq b$ and the first i for which $a_i \neq b_i$ has $a_i > b_i$, the finite-dimensional version of Hausner's proposition is: *there is a positive integer n and real-valued functions $u_1, \ldots, u_n$ on F such that, for all f, g $\in$ F and $\alpha \in [0, 1]$,*

$$fPg \quad \text{if and only if} \quad (u_1(f), \ldots, u_n(f)) \overset{L}{>} (u_1(g), \ldots, u_n(g)),$$
$$u_i(\alpha f+(1-\alpha)g) = \alpha u_i(f)+(1-\alpha)u_i(g) \quad \text{for} \quad i = 1, \ldots, n.$$

The generalization of this to arbitrary dimensionality involves the Well-Ordering Principle. Hausner's theory assumes that F is a 'mixture space', which is a mixture set that satisfies

M6. *If $\alpha, \beta \in [0, 1]$ and $\alpha\beta < 1$ then $\alpha[\beta f+(1-\beta)g]+(1-\alpha)h =$ $\alpha\beta f+(1-\alpha\beta)[\{\alpha(1-\beta)/(1-\alpha\beta)\}g+\{(1-\alpha)/(1-\alpha\beta)\}h]$.*

He uses axioms Al [weak order] and C2 [if $0 < \alpha < 1$ then fPg if and only if $\alpha f+(1-\alpha)hP\alpha g+(1-\alpha)h$] but deletes the Archimedean axiom A3. The absence of A3 gives rise to linear utility in more than one dimension.

Products of mixture sets

If $F_1, \ldots, F_n$ are mixture sets then so is $\prod_{i=1}^n F_i$ when $\alpha f+(1-\alpha)g$ is defined in the natural way as $(\alpha f_1+(1-\alpha)g_1, \ldots, \alpha f_n+(1-\alpha)g_n)$. In this case it is not hard to prove that: A1, A2 *and* A3 *for P on $\prod F_i$ imply that there are real-valued functions $u_1, \ldots, u_n$ on $F_1, \ldots, F_n$ respectively such that, for all $f, g \in \prod F_i$ and $f_i, g_i \in F_i$ and $\alpha \in [0, 1]$,*

$$fPg \quad \text{if and only if} \quad \textstyle\sum_{i=1}^n u_i(f_i) > \sum_{i=1}^n u_i(g_i),$$
$$u_i(\alpha f_i+(1-\alpha)g_i) = \alpha u_i(f_i)+(1-\alpha)u_i(g_i) \quad \text{for} \quad i = 1, \ldots, n.$$

Moreover, real-valued functions $v_1, \ldots, v_n$ satisfy these expressions in place of $u_1, \ldots, u_n$ if and only if there are numbers $a, b_1, \ldots, b_n$ with $a > 0$ such that, for each i, $v_i(f_i) = au_i(f_i)+b_i$ for all $f_i \in F_i$.

The proof of this theorem is based on the *NM* proposition. With u on $\prod F_i$ as given by that proposition we define $u_i(f_i)$ by $u_i(f_i) = u(f_1^0, \ldots, f_{i-1}^0, f_i, f_{i+1}^0, \ldots, f_n^0) - \sum_{j \neq i} u_j(f_j^0)$ where f^0 is fixed and the $u_i(f_i^0)$ satisfy $\sum u_i(f_i^0) = u(f^0)$. It is then shown that $\sum u_i(f_i) = u(f)$ for any $f \in \prod F_i$.

This development has an obvious interpretation when i refers to time and F_i is a set of probability measures on decision consequences in the ith time period. In a related development due to Fishburn (1965) and Pollak (1967), suppose that $\mathcal{P}$ is the set of all simple probability measures on the set of all subsets of a product set $X = \prod_{i=1}^n X_i$, and let p_i be the marginal measure of $p \in \mathcal{P}$ on X_i so that $p_i(x_i) = p(X_1 \times \ldots \times X_{i-1} \times \{x_i\} \times X_{i+1} \times \ldots \times X_n)$. Then if A1, A2 and A3 hold for P on $\mathcal{P}$ and if pIg whenever $p_i = q_i$ for $i = 1, \ldots, n$, it follows that there are real-valued functions $u_1, \ldots, u_n$ on $X_1, \ldots, X_n$ respectively such that, for all $p, q \in \mathcal{P}$,

$$pPq \quad \text{if and only if} \quad \sum_{i=1}^n E(u_i, p_i) > \sum_{i=1}^n E(u_i, q_i).$$

Some theory related to this is presented in Fishburn (1970, Chapter 11).

We conclude this chapter with a slight generalization of Theorem 13.3 in Fishburn (1970) that is a more general version of the theory in the final section of Chapter 9. We shall let $\mathcal{S}$ be the set of all subsets of a nonempty set S. F is assumed to be a mixture set with elements $f, g, \ldots,$ and $\mathcal{F}$ is the set of all functions on S to F. A typical element $\mathbf{f} \in \mathcal{F}$ assigns an $\mathbf{f}(s) \in F$ to each $s \in S$. On letting $(\alpha\mathbf{f}+(1-\alpha)\,\mathbf{g})(s) = \alpha\mathbf{f}(s)+(1-\alpha)\,\mathbf{g}(s)$ if follows that $\mathcal{F}$ is a mixture set. If S is finite with n elements, $\mathcal{F}$ is equivalent to the product $\prod F_i$ in the first paragraph of this section with $F_i = F$ for all i. The preference relation P will be applied to $\mathcal{F}$. $A \in \mathcal{S}$ is *null* if and only if $\mathbf{f}I\mathbf{g}$ whenever $\mathbf{f}(s) = \mathbf{g}(s)$ for every $s \in A^c$. We define $fP\mathbf{g}$ to mean that $\mathbf{f}P\mathbf{g}$ when $\mathbf{f}(s) = f$ for every $s \in S$. $\mathbf{f}Pg$ is defined in a similar way, and fPg if and only if $\mathbf{f}P\mathbf{g}$ when $\mathbf{f}(s) = f$ and $\mathbf{g}(s) = g$ for all $s \in S$. The following axioms are similar in intent to the six axioms in the final section of Chapter 9. The first three are A1, A2 and A3 for P on $\mathcal{F}$.

D1. *P on $\mathcal{F}$ is a weak order.*

D2. *If $\mathbf{f}P\mathbf{g}$ and $0 < \alpha < 1$ then $\alpha\mathbf{f}+(1-\alpha)\,\mathbf{h}P\alpha\mathbf{g}+(1-\alpha)\,\mathbf{h}$.*

D3. *If $\mathbf{f}P\mathbf{g}$ and $\mathbf{g}P\mathbf{h}$ then there are $\alpha, \beta \in\,]0, 1[$ such that $\alpha\mathbf{f}+(1-\alpha)\mathbf{h}P\mathbf{g}$ and $\mathbf{g}P\beta\mathbf{f}+(1-\beta)\,\mathbf{h}$.*

D4. *fPg for some $f, g \in F$.*
D5. *If $A \in \mathcal{S}$ is not null, if $\mathbf{f}(s) = f$ and $\mathbf{g}(s) = g$ for all $s \in A$, and if $\mathbf{f}(s) = \mathbf{g}(s)$ for all $s \in A^c$, then $\mathbf{f}P\mathbf{g}$ if and only if fPg.*
D6. *If $\mathbf{f}(s)\ P\mathbf{g}$ for all $s \in S$ or if $\mathbf{f}P\mathbf{g}(s)$ for all $s \in S$ then not $\mathbf{g}P\mathbf{f}$.*

These axioms imply that there is a real-valued function u on F and a probability measure p on $\mathcal{S}$ such that, with $u_{\mathbf{f}}(s) = u(\mathbf{f}(s))$ for all $s \in S$,

$$\mathbf{f}P\mathbf{g} \quad \text{if and only if} \quad E(u_{\mathbf{f}}, p) > E(u_{\mathbf{g}}, p),$$

for all $\mathbf{f}, \mathbf{g} \in \mathcal{F}$. Moreover, p is unique with $p(A) = 0$ if and only if A is null, and u is unique up to a positive linear transformation. Finally, u on F is bounded if there is a denumerable partition of S such that $p(A) > 0$ for every A in the partition, and if there is no such partition then there is a positive integer m such that no partition contains more than m events with positive probability.

CHAPTER 11

ordered groupoids

A *groupoid* is a nonempty set X and a function on $X \times X$ into X. The element in X assigned to $(x, y) \in X \times X$ by the function is often written as $x+y$ or xy or $x \bigcirc y$. I shall use $\bigcirc$.

Groupoids abound in mathematics. Some examples are: *Re* with $\bigcirc$ natural addition or multiplication; an algebra of sets or a topology $\mathcal{S}$ with $\bigcirc$ denoting union or intersection; a set $\mathcal{P}$ of probability measures on a given algebra with $\frac{1}{2}p+\frac{1}{2}q \in \mathcal{P}$ whenever $p, q \in \mathcal{P}$, in which case we can define $p \bigcirc q = \frac{1}{2}p+\frac{1}{2}q$. The operation $\bigcirc$ in each of these examples is *commutative* or Abelian since $x \bigcirc y = y \bigcirc x$ for all $x, y \in X$. In all but the last example $\bigcirc$ is *associative*, which means that $(x \bigcirc y) \bigcirc z = x \bigcirc (y \bigcirc z)$ for all $x, y, z \in X$. In the probability measures case we have $(p \bigcirc q) \bigcirc r = \frac{1}{4}p+\frac{1}{4}q+\frac{1}{2}r$ and $p \bigcirc (q \bigcirc r) = \frac{1}{2}p+\frac{1}{4}q+\frac{1}{4}r$ so that it is not generally true that $cp \bigcirc q) \bigcirc r = p \bigcirc (q \bigcirc r)$. An associative groupoid is a *semigroup*.

This chapter presents some basic theory of ordered groupoids. An *ordered groupoid* is a triple $(X, \bigcirc, R)$ consisting of a groupoid $(X, \bigcirc)$ and a transitive binary relation R on X that satisfies the following for all $x, y, z \in X$:

$$\text{if} \quad xRy \quad \text{then} \quad x \bigcirc z R y \bigcirc z \quad \text{and} \quad z \bigcirc x R z \bigcirc y.$$

In form, this is very much like the independence axiom of the preceding chapter: if fPg and $0 < \alpha < 1$ then $\alpha f+(1-\alpha) h P \alpha g+(1-\alpha) h$. Clearly, if $\bigcirc$ is commutative then $z \bigcirc x R z \bigcirc y$ is implied by $x \bigcirc z R y \bigcirc z$.

By prefixing various modifiers to 'ordered groupoids' and by replacing 'groupoids' with another term that subsumes the groupoid structure we can generate a rather large class of different types of ordered groupoids. For example, a partially ordered groupoid is an ordered groupoid for which R is a partial order, and a linearly ordered commutative semigroup is an ordered groupoid $(X, \bigcirc, P)$ for which P is a linear order and $(X, \bigcirc)$ is a commutative semigroup. There is a rather large mathematical literature for structures of this kind, a prime reference for which is Fuchs (1963). Other sources that I have drawn upon for this chapter include Aczél (1966), Pfanzagl (1968), Krantz (1968) and Vind (1969).

Ordered groups

One of the earliest real-valued order-preserving theorems for ordered groupoids is a theorem from Hölder (1901) for completely ordered groups. We shall consider this momentarily.

A *group* is a semigroup $(X, \bigcirc)$ for which there is an element $e \in X$, called the *group identity*, and for each $x \in X$ an element $x' \in X$ called the *inverse* of x such that, for all $x \in X$,

$$x \bigcirc e = e \bigcirc x = x$$

$$x \bigcirc x' = x' \bigcirc x = e.$$

$(Re, +)$ with $e = 0$ and $x' = -x$ is a group. $Re - \{0\}$ with $\bigcirc$ natural multiplication, $e = 1$, and $x' = 1/x$ is another group. Both groups are commutative or Abelian.

When $(X, \bigcirc, R)$ is an ordered group, xRy is equivalent to each of the following: $x \bigcirc y'Re$, $y' \bigcirc xRe$, $eRy \bigcirc x'$, $eRx' \bigcirc y$ and $y'Rx'$. For example, we have the following chain of implications: $xRy \Rightarrow x' \bigcirc xRx' \bigcirc y \Rightarrow eRx' \bigcirc y \Rightarrow e \bigcirc y'R(x' \bigcirc y) \bigcirc y' \Rightarrow y'Rx' \bigcirc (y \bigcirc y') \Rightarrow y'Rx' \bigcirc e \Rightarrow y'Rx' \Rightarrow y \bigcirc y'Ry \bigcirc x' \Rightarrow (y \bigcirc y') \bigcirc xR(y \bigcirc x') \bigcirc x \Rightarrow xRy$.

When $(X, \bigcirc)$ is a groupoid and m is a positive integer, we define $mx = x \bigcirc (x \bigcirc (\ldots \bigcirc (x \bigcirc x) \ldots))$. When $(X, \bigcirc)$ is a group, we define $0x = e$ and, for each negative integer m, $mx = x' \bigcirc (x' \bigcirc (\ldots \bigcirc (x' \bigcirc x') \ldots))$. If $(X, \bigcirc)$ is a group and m and n are integers then $mx \bigcirc nx = (m+n)x$.

Like Szpilrajn's theorem (1930) for extending a strict partial order to a linear order (or a partial order to a complete order), there are extension theorems for ordered groups. Fuchs (1963) discusses a number of these. Here is an example of such a theorem: *If $(X, \bigcirc, R)$ is a partially ordered commutative group then there is a completely ordered commutative group $(X, \bigcirc, R')$ with $R \subseteq R'$ if and only if $x = e$ whenever $x \in X$ and $mx = e$ for some positive integer m.* By our remarks in Chapter 3 and the group properties, a completely ordered group is essentially equivalent to a linearly ordered group.

If $(X, \bigcirc, R)$ is a completely ordered group, or if $(X, \bigcirc, P)$ is a linearly ordered group then (with $P = R \cap R^{c*}$ in the former case) it is *Archimedean* if and only if, for all $x, y \in X$, xPe and yPe imply that $mxPy$ for some positive integer m. Hölder's Theorem shows, among other things, that an Archimedean linearly ordered group is commutative.

HÖLDER'S THEOREM. *Suppose that $(X, \bigcirc, P)$ is a linearly ordered group. Then $(X, \bigcirc, P)$ is Archimedean if and only if there is a real-valued function f on such X that, for all $x, y \in X$,*

$$xPy \quad \text{if and only if} \quad f(x) > f(y),$$

$$f(x \bigcirc y) = f(x) + f(y).$$

Moreover, if both f and g have these properties then there is a real number $c > 0$ such that $g(x) = cf(x)$ for all $x \in X$ and c is unique (given f and g) if $\{e\} \subset X$.

Fuchs (1963, pp. 45–46) gives a proof of this. Essentially the same proof is repeated by Fishburn (1970, Section 5.1).

Krantz (1964) presents a proof of the Luce–Tukey theorem for two-factor additive measurement (1964) that is based on Hölder's Theorem. With a few minor changes the axioms of Luce and Tukey are: $X = X_1 \times X_2$, $I = (P \cup P^*)^c$ and, throughout X:

LT1. *If $x^k, y^k \in X$ for $k = 1, 2, 3$, if for each i ($i = 1, 2$) the sequence x_i^1, x_i^2, x_i^3 is a rearrangement of y_i^1, y_i^2, y_i^3, and if $x^1(P \cup I)y^1$ and $x^2(P \cup I)y^2$ then not x^3Py^3,*

LT2. *If $x_1, y_1 \in X_1$ and $x_2, y_2 \in X_2$ then there are $z_1 \in X_1$ and $z_2 \in X_2$ such that $(x_1, x_2)I(y_1, z_2)$ and $(x_1, x_2)I(z_1, y_2)$,*

LT3. *If $(x_1^0, x_2^1)P(x_1^0, x_2^0)$, if $(x_1^{k-1}, x_2^1)I(x_1^k, x_2^0)$ for $k = 1, 2, \ldots$ and if $y \in X$ then $(x_1^k, x_2^0)Py$ for some $k \in \{1, 2, \ldots\}$.*

Axiom LT1 implies that P on X is a weak order. Hence P' on X/I, defined by $aP'b$ if and only if xPy for some $x \in a$ and $y \in b$, is a linear order. With $(x_1^0, x_2^0) \in X$ fixed, we define $\bigcirc$ on X/I as follows:

$$a \bigcirc b \quad \text{is the equivalence class in } X/I \text{ that contains}$$

$$(x_1, x_2) \quad \text{when} \quad (x_1, x_2^0) \in a \quad \text{and} \quad (x_1^0, x_2) \in b.$$

With e such that $(x_1^0, x_2^0) \in e$ and a' the element in X/I that contains (x_1^0, x_2) when $(x_1, x_2^0) \in a$ and $(x_1, x_2) \in e$, it follows from LT1 and LT2 that $(X/I, \bigcirc, P')$ is a linearly ordered commutative group. Then LT3 is used to show that this group is Archimedean. Hölder's Theorem thus yields a real-valued function f on X/I that preserves P' and gives $f(a\bigcirc b) = f(a)+f(b)$. Defining $u_1(x_1) = f(a)$ when $(x_1, x_2^0) \in a$, and $u_2(x_2) = f(b)$ when $(x_1^0, x_2) \in b$, it follows that, for all $x, y \in X$,

$$xPy \quad \text{if and only if} \quad u_1(x_1)+u_2(x_2) > u_1(y_1)+u_2(y_2).$$

It should be mentioned that additive utilities can exist in this type of situation even when the group is not Archimedean. For example, let $X = \{(j, k)$: j and k are integers$\}$, let $\bigcirc$ be addition, and let P be the lexicographic order $\overset{L}{>}$ so that $(j, k)P(j', k')$ if and only if $j > j'$ or $j = j'$ & $k > k'$. Then $(X, +, P)$ is a linearly ordered commutative group with identity (0, 0), but it is not Archimedean since $(1, 0)P(0, 0)$ and $(0, 1)P(0, 0)$ but $m(0, 1) = (0, m)P(1, 0)$ for no positive integer m. However, there are real-valued functions u_1 and u_2 on the integers such that $(j, k)P(j', k')$ if and only if $u_1(j)+u_2(k) > u_1(j')+u_2(k')$. For example, $u_1(j) = j$ for all j and u_2 increasing in j with $\sup \{u_2(k)-u_2(j)\colon k > j\} < 1$ will suffice.

Over the years several theorems have evolved that give essentially the same conclusion as Hölder's Theorem but use weaker assumptions. One of the most recent of these is a theorem by Krantz (1968) for what he calls an Archimedean positive ordered local semigroup. For further discussion on this and related results, the reader is referred to Krantz (1968) and to Fuchs (1963).

Bisymmetric linearly ordered groupoids

A linearly ordered groupoid is of course an ordered groupoid $(X, \bigcirc, P)$ for which P is a linear order (transitive, irreflexive, weakly connected). In most cases the groupoids discussed in this section are not associative and hence are not semigroups. Intuitively this can be seen from the fact that we shall often interpret $x\bigcirc y$ as a element in X 'between' x and y. For example, with $x\bigcirc y = (x+y)/2$, $(Re, \bigcirc, >)$ is a linearly ordered groupoid that is not associative since $(x\bigcirc y)\bigcirc z = (x+y+2z)/4$ and $x\bigcirc(y\bigcirc z) = (2x+y+z)/4$.

An important axiom for ordered groupoids that goes back at least to Aczél (1948) and has been used by Fuchs (1950), Pfanzagl (1959, 1968) and Vind (1969) among others is the assumption of

BISYMMETRY. $(x \bigcirc y) \bigcirc (z \bigcirc w) = (x \bigcirc z) \bigcirc (y \bigcirc w)$ *for all* $x, y, z, w \in X$. This seems natural when $x \bigcirc y$ is viewed as the 'mean' of x and y or as a 'midpoint' between x and y. When $X = Re$, $[(x+y)/2+(z+w)/2]/2 = [(x+z)/2+(y+w)/2]/2$. We shall refer to a bisymmetric linearly ordered groupoid as a bisymmetric l.o.g.

Several interesting theorems for a bisymmetric l.o.g. have been developed by Aczél (1948, 1966), Fuchs (1950), Pfanzagl (1959, 1968) and Vind (1969). We consider two of these. The first is due to Aczél (1966, p. 287) and Pfanzagl (1968) and is in fact a special case of Pfanzagl's Theorem 6.1.1 (1968, p. 97).

THEOREM A. *Suppose that* $(X, \bigcirc, P)$ *is a bisymmetric* l.o.g. *and, with* $\mathcal{S}$ *the P-order topology for* X, $(X, \mathcal{S})$ *is connected and, for all* $a, b \in X$, *each of* $\{x\colon x \bigcirc aPb\}$, $\{x\colon a \bigcirc xPb\}$, $\{x\colon bPa \bigcirc x\}$ *and* $\{x\colon bPx \bigcirc a\}$ *is in* $\mathcal{S}$. *Then there is a continuous real-valued function* f *on* X *and numbers* α, β *and* γ *such that, for all* $x, y \in X$,

$$xPy \quad \text{if and only if} \quad f(x) > f(y),$$
$$f(x \bigcirc y) = \alpha f(x) + \beta f(y) + \gamma.$$

Moreover, f *is unique up to a positive linear transformation and if* X *contains more than one element then* α *and* β *are uniquely determined and are positive.*

Assume that X has more than one element and that f satisfies the theorem. Then if $(X, \bigcirc)$ is commutative it is easily seen that $\alpha = \beta$. If $(X, \bigcirc)$ is *idempotent*, which means that $x \bigcirc x = x$ for all $x \in X$, then $\alpha+\beta = 1$ and $\gamma = 0$. Finally, if $(X, \bigcirc)$ is both commutative and idempotent then $\alpha = \beta = \frac{1}{2}$ and $\gamma = 0$.

In this last case Vind (1969) proves a theorem that does not rely on topological properties and is more general than Theorem *A* for this case. With $(X, \bigcirc, P)$ a bisymmetric l.o.g., for each positive integer m let $(mx, z) = x \bigcirc (x \bigcirc \ldots \bigcirc (x \bigcirc z) \ldots)$ and $(x, mz) = (\ldots(x \bigcirc z) \bigcirc \ldots \bigcirc z) \bigcirc z$. We shall say that $(X, \bigcirc, P)$ is *Archimedean* if and only if $xPyPz$ implies that there is a positive integer m such that $(mx, z)PyP(x, mz)$. This property is similar to the Archimedean axiom A3 for mixture sets.

THEOREM B. *Suppose that* $(X, \bigcirc, P)$ *is an Archimedean bisymmetric* l.o.g. *and that* $(X, \bigcirc)$ *is commutative and idempotent. Then there is a real-valued function f on X such that, for all* $x, y \in X$,

$$xPy \quad \textit{if and only if} \quad f(x) > f(y),$$
$$f(x \bigcirc y) = \tfrac{1}{2}f(x) + \tfrac{1}{2}f(y),$$

and f is unique up to a positive linear transformation.

Both Pfanzagl (1968) and Vind (1969) demonstrate the relevance of theorems like *A* and *B* for a number of areas in the theory of measurement, including decision-theoretic subjects such as additive utilities, comparable preference differences, and expected utility. Several theorems that are presented in decision-theoretic settings and are more or less similar to Theorem *B* are presented by Debreu (1959, 1960), Suppes and Winet (1955), Scott and Suppes (1958), Suppes and Zinnes (1963), Chipman (1960), and Luce (1968).

We conclude with an application of Theorem *B* in the area of comparable preference differences. We begin by assuming that $(X, \bigcirc, P)$ is a weakly ordered groupoid and interpret $x \bigcirc y$ as an element in X that is midway in preference between x and y. The judgment of a preferential midpoint is of course a stronger judgment than is presupposed by ordinary weak order. A special relational symbol (introduced below) is often used for the notion of comparable differences in preference. The formulation used here incorporates this special preference notion within the interpretation of $\bigcirc$.

With $I = (P \cup P^*)^c$ we shall, as usual, let X/I be the set of equivalence classes of X under I and take $aP'b$ if and only if xPy for some $x \in a$ & $y \in b$. P' on X/I is a linear order.

In addition to our initial assumptions for $(X, \bigcirc, P)$ we use four assumptions pertaining to I: for all $x, y, z, w \in X$,

1. xIy implies $x \bigcirc z I y \bigcirc z$,
2. $(x \bigcirc y) \bigcirc (z \bigcirc w) I (x \bigcirc z) \bigcirc (y \bigcirc w)$,
3. $x \bigcirc y I y \bigcirc x$,
4. $x \bigcirc x I x$.

These have obvious interpretations in terms of our interpretation of $\bigcirc$ and the preceding discussion in this chapter. With $a \bigcirc' b$ the equivalence class in X/I that contains $x \bigcirc y$ when $x \in a$ & $y \in b$, it is not hard to show that our assumptions imply that $(X/I, \bigcirc', P')$ is a bisymmetric l.o.g. and that

$(X/I, \bigcirc')$ is commutative and idempotent. Hence, under the appropriate Archimedean assumption there is a real-valued function f on X/I such that $f(a) > f(b)$ if and only if $aP'b$, and $f(a\bigcirc' b) = f(a)/2 + f(b)/2$. Defining u on X from f on X/I in the natural way we get $u(x) > u(y)$ if and only if xPy, along with $u(x \bigcirc y) = u(x)/2 + u(y)/2$.

Under our interpretation of $\bigcirc$ we define a binary relation P^0 on $X \times X$ as: $(x, y)P^0(z, w)$ if and only if $x\bigcirc zPy\bigcirc w$. The interpretation of $(x, y)P^0(z, w)$ is that the degree of preference difference from y to x exceeds the degree of preference difference from w to z. This is a directed notion. Under the properties for u in the preceding paragraph it follows that, for all $x, y, z, w \in X$,

$$(x, y)\, P^0(z, w) \quad \text{if and only if} \quad u(x) - u(y) > u(z) - u(w).$$

references

Aczél, J. (1948), 'On mean values', *Bull. Amer. Math. Soc.* 54, 392–400.

Aczél, J. (1966), *Lectures on functional equations and their applications.* New York: Academic Press.

Adams, E. W. (1965), 'Elements of a theory of inexact measurement', *Phil. Sci.* 32, 205–228.

Armstrong, W. E. (1939), 'The determinateness of the utility function', *Econ. J.* 49, 453–467.

Arrow, K. J. (1950), 'A difficulty in the concept of social welfare', *J. Polit. Econ.* 58, 328–346.

Arrow, K. J. (1959), 'Rational choice functions and orderings', *Economica* 26, 121–127.

Arrow, K. J. (1963), *Social choice and individual values*, 2nd. Ed. New York: Wiley.

Aumann, R. J. (1962), 'Utility theory without the completeness axiom', *Econometrica* 30, 445–462. Correction: *Econometrica* 32, 210–212.

Berge, C. (1958), *Théorie des graphes et ses applications.* Paris: Dunod. [English translation (1962). New York: Wiley.]

Birkhoff, G. (1948), *Lattice theory.* New York: American Mathematical Society.

Blackwell, D. and Girshick, M. A. (1954), *Theory of games and statistical decisions.* New York: Wiley.

Blau, J. H. (1957), 'The existence of social welfare functions', *Econometrica* 25, 302–313.

Bowen, R. (1968), 'A new proof of a theorem in utility theory', *International Econ. Rev.* 9, 374.

Cantor, G. (1895), 'Beiträge zur Begründung der transfiniten Mengenlehre', *Math. Ann.* 46, 481–512.

Cantor, G. (1915), *Contributions to the founding of the theory of transfinite numbers.* Trans. with notes by P.E.B. Jourdain. New York: Dover.

Chipman, J. S. (1960), 'The foundations of utility', *Econometrica* 28, 193–224.

Chipman, J. S. (1971), 'Consumption theory without transitive indifference', in *Preferences, utility and demand*, ed. by J. S. Chipman, L. Hurwicz, M. K. Richter and H. Sonnenschein. New York: Harcourt Brace Jovanovich.

Cohen, P. J. (1966), *Set theory and the continuum hypothesis*. New York: Benjamin.

Debreu, G. (1954), 'Representation of a preference ordering by a numerical function', in *Decision processes*, ed. by R. M. Thrall, C. H. Coombs and R. L. Davis. New York: Wiley.

Debreu, G. (1959), 'Cardinal utility for even-chance mixtures of pairs of sure prospects', *Rev. Econ. Studies* 26, 174–177.

Debreu, G. (1960), 'Topological methods in cardinal utility theory', *In Mathematical methods in the social sciences, 1959*, ed. by K. J. Arrow, S. Karlin and P. Suppes. Stanford, California: Stanford University Press.

Debreu, G. (1964), 'Continuity properties of Paretian utility', *International Econ. Rev.* 5, 285–293.

de Finetti, B. (1937), 'La prévision: ses lois logiques, ses sources subjectives', *Ann. de l'Institut Henri Poincaré* 7, 1–68 [English translation in Kyburg and Smokler (1964).]

Domotor, Z. (1969), 'Probabilistic relational structures and their applications', Tech. Report 144, *Institute for Mathematical Studies in the Social Sciences*, Stanford University. Stanford, California.

Eilenberg, S. (1941), 'Ordered topological spaces'. *Amer. J. Math.* 63, 39–45.

Fishburn, P. C. (1965), 'Independence in utility theory with whole product sets', *Operations Research* 13, 28–45.

Fishburn, P. C. (1969), 'Weak qualitative probability on finite sets', *Ann. Math. Statist.* 40, 2118–2126.

Fishburn, P. C. (1970), *Utility theory for decision making*. New York: Wiley.

Fuchs, L. (1950), 'On mean systems', *Acta Math. Acad. Sci. Hung.* 1, 303–320.

Fuchs, L. (1963), *Partially ordered algebraic systems*. London: Pergamon Press. Reading, Massachusetts: Addison-Wesley.

Halphen, E. (1955), 'La notion de vraisemblance', *Pub. de l'Institut de Statist. de l'Université de Paris* 4, 41–92.

Hansson, B. (1968), 'Choice structures and preference relations', *Synthese* 18, 443–458.

Harary, F., Norman, R. Z. and Cartwright, D. (1965), *Structural models: an introduction to the theory of directed graphs*. New York: Wiley.

Hausner, M. (1954), 'Multidimensional utilities', in *Decision processes*, ed. by R. M. Thrall, C. H. Coombs and R. L. Davis. New York: Wiley.

Herstein, I. N. and Milnor, J. (1953), 'An axiomatic approach to measurable utility', *Econometrica* 21, 291–297.

Hölder, O. (1901), 'Die Axiome der Quantität und die Lehre vom Mass', *Berichte Verhand. König. Sächs. Gesell. Wiss.* (Leipzig), *Math. Phys. Kl.* 53, 1–64.

Houthakker, H. S. (1950), 'Revealed preference and the utility function', *Economica* 17, 159–174.

Houthakker, H. S. (1961), 'The present state of consumption theory', *Econometrica* 29, 704–740.

Hurwicz, L. and Richter, M. K. (1971), 'Revealed preference without demand continuity assumptions', in *Preferences, utility and demand,* ed. by J. S. Chipman, L. Hurwicz, M. K. Richter and H. Sonnenschein. New York: Harcourt Brace Jovanovich.

Jensen, N. E. (1967), 'An introduction to Bernoullian utility theory. I. Utility functions', *Swedish J. Econ.* 69, 163–183.

Kannai, Y. (1963), 'Existence of a utility in infinite dimensional partially ordered spaces', *Israel J. Math.* 1, 229–234.

Kelley, J. L. (1955), *General topology.* Princeton, New Jersey: Van Nostrand.

Klee, V. L. (1955), 'Separation properties of convex cones', *Proc. Amer. Math. Soc.* 6, 313–318.

Koopman, B. O. (1940), 'The axioms and algebra of intuitive probability', *Ann. Math.* 41, 269–292.

Koopman, B. O. (1941), 'Intuitive probabilities and sequences', *Ann. Math.* 42, 169–187.

Kraft, C. H., Pratt, J. W. and Seidenberg, A. (1959), 'Intuitive probability on finite sets', *Ann. Math. Statist.* 30, 408–419.

Krantz, D. H. (1964), 'Conjoint measurement: the Luce–Tukey axiomatization and some extensions', *J. Math. Psychol.* 1, 248–277.

Krantz, D. H. (1968), 'A survey of measurement theory', in *Mathematics of decision sciences,* part 2, ed. by G. B. Dantzig and A. F. Veinott. Providence, Rhode Island: American Mathematical Society.

Kuhn, H. W. and Tucker, A. W. (Eds.) (1956), *Linear inequalities and related systems.* Princeton, New Jersey: Princeton University Press.

Kyburg, H. E. and Smokler, H. E. (Eds.) (1964), *Studies in subjective probability.* New York: Wiley.

Loève, M. (1963), *Probability theory,* 3rd Ed. Princeton, New Jersey: Van Nostrand.

Luce, R. D. (1956), 'Semiorders and a theory of utility discrimination', *Econometrica* 24, 178–191.

Luce, R. D. (1967), 'Sufficient conditions for the existence of a finitely additive probability measure', *Ann. Math. Statist.* 38, 780–786.

Luce, R. D. (1968), 'On the numerical representation of qualitative conditional probability', *Ann. Math. Statist.* 39, 481–491.

Luce, R. D. and Suppes, P. (1965), 'Preference, utility, and subjective probability', in *Handbook of Mathematical Psychology,* Vol. III, ed. by R. D. Luce, R. R. Bush and E. Galanter. New York: Wiley.

Luce, R. D. and Tukey, J. W. (1964), 'Simultaneous conjoint measurement: a new type of fundamental measurement', *J. Math. Psychol.* 1, 1–27.

May, K. O. (1952), 'A set of independent necessary and sufficient conditions for simple majority decision', *Econometrica* 20, 680–684.

May, K. O. (1953), 'A note on the complete independence of the conditions for simple majority decision', *Econometrica* 21, 172–173.

Mirsky, L. and Perfect, H. (1966), 'Systems of representatives', *J. Math. Analysis and Applications* 15, 520–568.

Nachbin, L. (1965), *Topology and order*. Princeton, New Jersey: Van Nostrand.

Newman, P. and Read, R. (1961), 'Representation problems for preference orderings', *J. Econ. Behavior* 1, 149–169.

Ore, O. (1962). *Theory of graphs*. Providence, Rhode Island: American Mathematical Society.

Pfanzagl, J. (1959), 'A general theory of measurement: applications to utility', *Naval Research Logistics Quart.* 6, 283–294.

Pfanzagl, J. (1968), *Theory of measurement*. New York: Wiley.

Pollak, R. A. (1967), 'Additive von Neumann–Morgenstern utility functions', *Econometrica* 35, 485–494.

Rader, J. T. (1963), 'The existence of a utility function to represent preferences', *Rev. Econ. Studies* 30, 229–232.

Richter, M. K. (1966), 'Revealed preference theory', *Econometrica* 34, 635–645.

Samuelson, P. A. (1938), 'A note on the pure theory of consumer's behaviour', *Economica* 5, 61–71 and 353–354.

Samuelson, P. A. (1948), 'Consumption theory in terms of revealed preference', *Economica* 15, 243–253.

Savage, L. J. (1954), *The foundations of statistics*. New York: Wiley.

Scott, D. (1964), 'Measurement structures and linear inequalities', *J. Math. Psychol.* 1, 233–247.

Scott, D. and Suppes, P. (1958), 'Foundational aspects of theories of measurement', *J. Symbolic Logic* 23, 113–128.

Sonnenschein, H. (1965), 'The relationship between transitive preferences and the structure of the choice space', *Econometrica* 33, 624–634.

Suppes, P. and Winet, M. (1955), 'An axiomatization of utility based on the notion of utility differences', *Management Sci.* 1, 259–270.

Suppes, P. and Zinnes, J. L. (1963), 'Basic measurement theory', in *Handbook of Mathematical Psychology*, Vol. I, ed. by R. D. Luce, R. R. Bush and E. Galanter. New York: Wiley.

Szpilrajn, E. (1930), 'Sur l'extension de l'ordre partiel', *Fund. Math.* 16, 386–389.

Tversky, A. (1964), 'Finite additive structures', *Michigan Mathematical Psychology Program*, M M P P 64–6. University of Michigan, Ann Arbor, Michigan.

Uzawa, H. (1960), 'Preference and rational choice in the theory of consumption', in *Mathematical methods in the social sciences, 1959*, ed. by K. J. Arrow, S. Karlin and P. Suppes. Stanford, California: Stanford University Press.

Vind, K. (1969), *Mean groupoids*. Copenhagen.

Villegas, C. (1964), 'On qualitative probability σ-algebras', *Ann. Math. Statist.* 35, 1787–1796.

Villegas, C. (1967), 'On qualitative probability', *Amer. Math. Monthly* 74, 661–669.

von Neumann, J. and Morgenstern, O. (1947), *Theory of games and economic behavior*, 2nd Ed. Princeton, New Jersey: Princeton University Press.

Wold, H. (1943), 'A synthesis of pure demand analysis: I, II, III', *Skandinavisk Aktuarietidskrift* 26.

index

Bei Fragen zur Produktsicherheit wenden Sie sich bitte an:
If you have any questions regarding product safety,
please contact:

Walter de Gruyter GmbH
Genthiner Straße 13
10785 Berlin
productsafety@degruyterbrill.com